Walter Angus Elmslie

Among the Wild Ngoni

Some Chapters in the History of the Livingstonia Mission in British Central Africa

Walter Angus Elmslie

Among the Wild Ngoni
Some Chapters in the History of the Livingstonia Mission in British Central Africa

ISBN/EAN: 9783744762502

Printed in Europe, USA, Canada, Australia, Japan

Cover: Foto ©berggeist007 / pixelio.de

More available books at **www.hansebooks.com**

MRS. ELMSLIE.
DR. ELMSLIE.

AMONG THE WILD NGONI

BEING SOME CHAPTERS IN THE HISTORY
OF THE LIVINGSTONIA MISSION IN
BRITISH CENTRAL AFRICA

BY

W. A. ELMSLIE
M.B., C.M., F.R.G.S.
Medical Missionary

WITH INTRODUCTION BY
THE RIGHT HON. LORD OVERTOUN

FLEMING H. REVELL COMPANY
NEW YORK   CHICAGO   TORONTO
Publishers of Evangelical Literature
1899

To

MY WIFE

A TRUE HELPMEET IN ALL MY WORK

CONTENTS

ILLUSTRATIONS

INTRODUCTION

THE eyes of the world are on Africa, and the nations of the West are eagerly engaged in exploring and annexing land without asking the consent of the inhabitants. Till far on in the century only the fringes of Africa were known, the districts round the Cape up to Natal were early colonised, while the West Coast was specially known as the "White Man's Grave." The north, once the abode of pirates, fell chiefly under French influence, and the wondrous land of Egypt, stretching into the dim past, has been the battlefield of hosts contending for its possession. While the East Coast has languished under Portuguese misrule and neglect, Egypt and the southern regions have steadily advanced under British possession and influence.

The southern portion of what has long been known as the Dark Continent has been to a great extent civilised, and while elements have not been wanting to degrade the native races, much has been done to spread the Gospel and the arts of peace. But during all these years the interior of Africa was an unknown land, sometimes marked in maps as "Desert," but believed to be the abode of horrid cruelty. Explorers from Bruce to Speke, Thomson and Grant, sought to penetrate its secrets, but the malarial climate, the fever swamps and tangled forests, not to speak of wild beasts and savage men, barred the way.

It was David Livingstone, a self-educated Scottish weaver, who, inspired with the passion to discover the secret sources of the Nile, and the mysteries of Central Africa, was raised up by God to carry the Gospel message to those who, for centuries, had sat in darkness and in the shadow of death.

This is not the place to recite how, time after time,

he plunged alone into the dark land, and with a gentleness which won his way, and a dauntless and persevering daring which carried him through many perils, brought to light the secrets of centuries, and blazed a path for civilisation and the Gospel.

But his heart was wrung with the horrors of the dreadful slave trade which had decimated Africa for ages, and caused the groans and sighs of her sons and daughters to ascend to heaven.

On a May day in 1873, worn out by fatigue and cruel fever, he was found dead by his faithful native boys, kneeling as in prayer at the side of the rude bed in his hut, amid the swamps of Lake Bangweolo.

Among his last written words were, " May Heaven's rich blessing come down on every one—American, English, Turk—who will help to heal this open sore of the world."

Carried by loving hands over a nine months' march, his body was laid in Westminster Abbey in April 1874, and the story of his life and death sent a thrill through Christendom, and purposes were formed for the sending of the Gospel to Central Africa.

Dr James Stewart of Lovedale was the first to move, and the result was the formation of the Livingstonia Mission by the Free Church of Scotland, the Blantyre Mission by the Church of Scotland, the Universities' Mission by the Universities of Oxford and Cambridge, and the Tanganyika Mission by the London Missionary Society, while later Moravians, Germans, and others followed.

The sphere chosen by the Livingstonia Mission was the west shore of Lake Nyasa, an inland sea some 400 miles long, discovered by Livingstone and Dr Stewart; and in 1875 the *Ilala*, bearing the pioneers of the Mission, Dr Laws and his helpers, steamed into the Lake and took possession for Christ.

The first settlement at Cape Maclear, the south end of

the Lake, had to be abandoned because of its unhealthy climate, which cost the lives of several missionaries. Moving to Bandawe, about half way up the West Coast of Lake Nyasa, the pioneers settled there, surveyed the land around, and began its conquest for the Gospel.

The story of the Livingstonia Mission is one of faithful and persevering work in the face of untold difficulties. Unknown languages had to be mastered and reduced to writing. Slavery and barbarism faced the Missionaries at every point. An unknown tropical climate tried them to the uttermost. When one fell at his post another stepped into the breach. Supported by prayer, faith and patience, they laboured on for years, till at last the seed sown in tears took root and sprang up. Now the labourers are filled with praise because God has given them to see fields white to the harvest.

The work has been carried on all these years by men and women, whose names shine as heroes in the Gospel story, on four great lines:—

1. The direct proclamation of the Gospel.
2. Education of young and old.
3. Medical Mission work.
4. Industrial training.

These have all been carried on at each of five stations, which all have many out-stations. In recent years a great central Training Institution has been established at Kondowi, to which the best pupils are drafted to be trained as evangelists, teachers, and skilled artisans. There are now some 500 resident and day students, and Dr Laws, who has been the honoured head of the Mission since its beginning, is in charge.

The Livingstonia Mission seeks to evangelise a field of about 300 miles long by 100 miles broad. There are now 7 native churches with over 1000 members, 85 schools with 11,000 scholars, and 300 native teachers and preachers.

While the whole field is full of the deepest interest, and each tribe has its own character, traditions and peculiarities, one of the tribes is dominant. The Ngoni, of whom the following pages tell, are the warriors of the country, of Zulu race, with splendid physique and qualities, but steeped for centuries in superstition, bloodshed, and cruelty. The fascinating story told by Dr Elmslie of the rise of Chaka's kingdom, of the seas of blood shed by him and his warriors, accompanied by untold cruelties, and all for lack of the Gospel unsent by sleeping Christendom, should stir the hearts of many to send the message of peace where it has not yet gone.

Dr Elmslie, who with his devoted wife has just sailed for Africa to begin his third term of service, vividly pictures the lofty plateau of Ngoniland, with its native villages and the dark background of vice and cruelty which lies behind the village life, with the horrors of the slave trade which harried peaceful homes, leaving the smoking ruins, while the inmates were massacred, or reserved for a more cruel fate, and how their perils drove the people to live in swamps or inaccessible rocks.

The first advance of the missionaries to Ngoniland was in 1878 in the face of much personal danger. The first interviews with Mombera and his bloodthirsty chiefs, picture not only the danger of the situation, but the faith, courage, and tact of the men who, taking their lives in their hands, went as ambassadors of Christ to these bloodstained savages.

They were worth winning for Christ, but it was a long story of alternating hope and fear, of patience and trial. The inquisitive questions, the insatiable and insolent greed shown to the missionaries, who were known not only to have brought "The Book," but calico and beads, were most trying.

The story of William Koyi, a Kafir Christian trained at Lovedale, and how, with Christian tact and patience, he disarmed suspicion, and secured for himself and his Europeans the friendship of Mombera and his people, has seldom been equalled in the missionary field.

To preach to the people was at first well-nigh impossible, the time of sowing had not yet come, much less the reaping; but the influence of his humble Christian life and example in the face of danger and difficulty, won at last the respect and love of the Ngoni tribe. Dr Elmslie touchingly tells how William Koyi, the faithful worker, heard on his dying bed that full permission had been given to teach and preach the Gospel, and with "nunc dimittis" on his lips went to his reward in 1886.

For full three years the pioneers laboured, prayed, and watched. The medical aid given helped them to win their way among the people, who wondered why they remained when no one would receive their message.

There came the first tiny blade when three youths came like Nicodemus at night to inquire, but these first-fruits met with bitter opposition, and Dr Elmslie and his faithful helpers were sorely tried by dangers, anxieties, fever, and disappointment. Then came the turning-point when, after a long drought, rain fell in response to the white men's prayers, and a new era began.

Mrs Elmslie's arrival created a fresh interest; work was begun among the girls, as had been done by Mrs Laws at Bandawe, and after a while, on the people's own proposal, they had a harvest thanksgiving to God.

Dr Elmslie tells the life-story of James Sutherland of Wick, converted in connection with D. L. Moody's mission there, who faithfully laboured with the Doctor amid dangers and difficulties, and who, before his death, showed such enthusiasm that when, in consequence of murderous threats, plans were made for the missionaries leaving, Sutherland had arranged to become a slave to one of the Ngoni in order to remain as a witness for God among the people.

The story of the exorcising of spirits, of Dr Laws' visit,

and the terrible suspense which the missionaries passed through, lead up to the first baptism in 1890. Then Dr Steele began his too brief work, which for five years brightened the band of workers, till his valued life was laid down.

In 1892 the first Ngoni woman was baptised ; two years later Miss Stewart joined the workers, and that year 760 children attended school.

Then the most northern station was opened at Mwenzo by Mr and Mrs Dewar and the Training Institution was started at Kondowi, above Florence Bay.

While Europeans must be pioneers (and God has given the Livingstonia Mission a splendid staff), the evangelisation of Africa must be done by Africa's sons, and the 500 students in training at the Institution who will soon be the craftsmen, teachers, evangelists, and pastors of British Central Africa.

The Rev. Donald Fraser, who has been nearly a year in Ngoniland, has had the joy of helping the earlier labourers in the reaping of the harvest which now gladdens the hearts of all.   At Ekwendeni he joined the Rev. James Henderson and others in a great Communion service when 195 sat at the Lord's Table, in presence of 4000 natives. In two days 198 adults and 89 children were baptised.

The scenes so graphically described in these pages, of warriors who once marched in impis to bloodshed and cruelty, now marching in hundreds to a Gospel gathering, witnessing the sacraments of the Lord's Supper and Baptism with reverent interest ; of the night air vocal with hymns where once the war-cry was heard; of peaceful homes and cultivated land, all tell of the triumph of the Gospel of God, and how, through the labours of Dr Laws, Dr Elmslie and their noble band as well as those who have gone to their rest, the wilderness and the solitary place is glad for them and the desert rejoices and blossoms as the rose.

OVERTOUN.

CHAPTER I

THE HISTORY OF THE NGONI

IN order to understand the present character of
the Ngoni it is necessary to go back to the
dawn of the present century and to South
Africa, the cradle of these people. The mighty
movements of barbarous fanatics in recent times,
such as those in the Soudan and elsewhere, sink
into insignificance when compared with those
that give rise to the presence of the Ngoni in
British Central Africa and in German East
Africa, not to speak of the Matabele who gave
so much trouble to the British, or the other
branches of the same race which had to be
proceeded against by Portuguese arms.

In a district somewhere on, or near, the Tugela
river, which now forms the northern boundary
of the colony of Natal, there was born, as the
century dawned, a child with a reputed miracu-
lous origin but fathered by Senzangakona, chief
of the then insignificant Zulu tribe. His mother,
fearing for his life, fled with him to the court

of a neighbouring and more powerful chief, named Dingiswayo, ruling the Amatetwa. Here he was received and cared for until he attained to manhood. Umnandi (*i.e.* the pleasant one) the mother of Chaka, as the child was named, remained with him. Dingiswayo was at that time the most powerful chief in the district stretching from Natal to Delagoa Bay. He had, in the early part of his life, been compelled to flee into what is now part of Cape Colony, and while exiled is supposed to have come into some knowledge of carrying on war by organised regiments and companies.

Thus through Europeans came the impulse which, as we shall see, was destined to have such awful results in the life and history of individuals and tribes over nearly half the length of Africa. On gaining the chieftainship of his tribe Dingiswayo organised his army in regiments, and otherwise improved its means of carrying desolation over a wide area. Chaka, during his stay with Dingiswayo, had no doubt ample opportunity for studying the art of war and seems to have done so successfully. He even improved on Dingiswayo's methods, and was not satisfied that conquered tribes should be so generously treated as they were by being incorporated as vassals of the paramount chief. Chaka saw in this a source of insecurity and formed the idea of so disorganising or crushing them that they would be incapable of rising against the chief. No doubt it was his education in war and bloodshed that bore its fruits when, as a young man, he ascended the throne of his father by causing the death of his brother, to supplant whom he had returned to the Zulu tribe against the wish of Dingiswayo. As Dingiswayo had opposed his pretensions to rule he had him " removed " soon after.

With this Chaka our history of the Ngoni begins. His brief reign of seven or eight years was a period in which more blood was shed, and greater upheaval among native tribes induced, than in any other country in the world. As a writer says, " War poisoned all enjoyment, cut off all that sustains life, turned thousands of square miles into literally a howling wilderness, shed rivers of blood, annihilated whole communities, turned the members of others into cannibals, and caused miseries and sufferings, the full extent of which can never now be known, and which, if ever known, could not be told." These words were written on the death of Chaka in 1828, and although it is estimated that over a million human beings owed their death either directly or indirectly to Chaka, it is not im-

probable that over the region which the wave of
war and bloodshed travelled, even more than
that number were slain in battle, massacred in
their villages, or driven into the wilds to die
of starvation.

The last of the chiefs to be conquered by
Chaka was Zwide, under whom Zongandaba and
other chiefs (formerly independent) ruled over
districts and acted as commanders of divisions
of his army. In a great battle with Zwide and
all his chiefs Chaka was victorious, as the proud
Ngoni are careful to state, through the deception
of a man named Noluju, who was a political
prisoner with Zwide and who desired to "pay
out" Zwide for some wrong done him. This
Noluju went to Chaka and arranged that, on the
attack being made, he would mislead the army
of Zwide. Arranging that Chaka's army should
camp by some favourable watering-place, he
guided Zwide's force to a barren place and left
it there, under pretence of going to spy out
Chaka's position. When they were faint from
thirst he guided them to where the water was,
but Chaka attacked them, killing many and
putting the rest to flight. The different chiefs
who had thus been united under Zwide again
sought independence by leaving the country,
and the Ngoni who are now in British Central

Africa then began their wanderings, every step of which is marked by blood as we shall see. Of Noluju it is related that, having returned home to get his wives, he set out for another place in which to live. Lying at night in his booth in the forest and evidently congratulating himself on having paid out Zwide for his treatment of him, he put his thoughts in song, native-wise, and sang, " He forgets who did the wrong, but he forgets not who was wronged." Zwide's spies, sent to chase him, heard the song and fell on him and killed them all. The life and fate of this unit illustrates the life and fate of many tribes. Noluju's song was a pæan and a prophecy, and he himself the subject.

Although the Ngoni lived under Zwide they were not in entire subjection to him, and on occasion, as their own tributaries have done since, they rose in rebellion. As illustrating how, even in those dark days, right principle was not without a witness, and was found in the heart of a woman, and how the superstitions of the people enter into and influence every act of their life, the following native narrative may be given.

Zwide, who had attacked Zongandaba, was taken prisoner, and on being released after some months was sent home under escort with a gift of many cattle. His pride was wounded by this insult from one of his vassals, and he determined on revenge. His mother opposed it, but to her he would not give heed. She devised a plan to strike fear into the hearts of the soldiers. In the words of a native, it is stated that " his mother reasoned with him, saying, ' My child, shall the Ngoni perish ? Did they not send you back, giving you many fat cattle with you ? Is it right to go out to war against them ? ' But Zwide gave no heed to his mother's words, and called together his soldiers. On the day when they were being reviewed, the mother of Zwide, having planned to make the soldiers afraid, went into the cattle-fold (it was not permissible for women to do so) where the soldiers were. Standing in their midst she unloosed her skirt and stood exposed among them. The soldiers seeing her thus wondered greatly, and Zwide also wondered. The soldiers declared that it was an omen, that perhaps an ancestral spirit had prompted her to do thus, and they, being afraid to go out, were disbanded forthwith. So Lowawa, Zwide's mother, prevailed."

On the breaking up of Zwide's combined force, Zongandaba and other petty chiefs led off sections of the tribe in quest of new lands, as they could not retain their old country against the growing power of Chaka. They had been conquered, but they had evidently been impressed by Chaka's methods, and resolved to follow them. No doubt also they appropriated the fame of Chaka and would be looked upon with fear by the weaker tribes they resolved on attacking. They passed through the Swazi country, attacking the people, impressing many to join them and capturing many cattle. Not many of the Swazi tribe lived to settle with the Ngoni west of Nyasa, but the oldest person in the country, probably, is a Swazi woman whose husband was a contemporary of Zongandaba, and afterwards a sub-chief of Mombera's. Having increased their strength and wealth by this attack on the Swazis, the horde then entered Tongaland to the west of Delagoa Bay, and settled for a time on the lower reaches of the Limpopo river. They crossed the Nkomati river near where there is now a station of the Basel Mission. Here a petty chief of Chaka, named Nqaba, with a following came upon them and there was a battle. Nqaba was driven back, but Zongandaba did not feel safe even there from an attack by Chaka. Having added to their force and their wealth by annexing many Tonga and their cattle, they went towards the west and attacked the Karanga tribe. Here, as among the Tonga, they instructed them in their methods of warfare and were gaining in

power by these additions.   After a short residence among the Karanga another move was
made towards the north, and they arrived in
June 1825 at the Zambezi somewhere between
Zumbo and Tette.

Here it may be of interest to turn aside and
complete our narrative of the waves of bloodshed
set rolling by Chaka, by glancing at the rise of
two kingdoms south of the Zambezi, under two
chiefs driven out by him about the time that the
Ngoni began their wanderings.   The first is that
of Gazaland, first occupied by Sotshangane who
fought with Zongandaba under Zwide against
Chaka, and fled at the same time.   We may safely
infer that his progress northward was marked by
blood, and that he and his successor Umzila did
not organise their vast kingdom, before then
composed of many small tribes, without much
more bloodshed.   But who can tell what suffering
and death resulted ?   When Umzila died, his son
Gungunhana succeeded, and in the recent opening up of Africa he has given as much trouble to
the Portuguese as the Matabele have given to
the British.

The other great power for evil springing up
at this time was Umziligazi, who fled from the
tyranny of Chaka and settled in the north of the
Transvaal.   His name inspired terror through

a vast region, as he completely subjugated or destroyed every tribe from whose opposition he had anything to dread. Readers of " Robert and Mary Moffat" will remember that this is "the scourge of the Bechuanas," "the Napoleon of South Africa," to whom Dr Moffat went first in 1829. Afterwards, when he had removed further north, Dr Moffat travelled 700 miles to see him and seek his salvation. Umziligazi formed a strong attachment to Dr Moffat, which was continued for thirty-nine years, until he died in 1868. The accounts given by Dr Moffat of these visits should be read by every one, but I cannot help quoting from his biography by his son, referred to above. It describes Dr Moffat's farewell to the great chief in 1860, when the veteran laid down his work at Inyati where the Mission had been planted. " On Sunday morning, the 17th June, he walked up to the chief's kraal, for the purpose of speaking to Umziligazi and his people for the last time on the great themes of life, death, and eternity. As we followed him along the narrow path, from our camp to the town about a mile distant, winding through fields and around patches of uncleared primeval forest, no step was more elastic and no frame more upright than his. In spite of unceasing toil and tropical heats and miasmatic exhalations, in spite of cares and disappointments, his wonderful energy seemed unabated. The old chief was as usual in his large court-yard, and gave kindly greeting. They were a strange contrast as they sat side by side—the Matabele tyrant and his friend the messenger of peace. The word of command was given ; the warriors filed in and arranged themselves in a great semicircle, sitting on the ground, the women crept as near as they could, behind huts and other points of concealment, and all listened in breathless silence to the last words of ' Moshete.' He himself knew that they were his last words, and that his work in Matabeleland was now given over to younger hands. It was a solemn service, and closed the long series of such, in which the friend of Umziligazi had striven to pierce the dense darkness of soul which covered him and his people. On the morrow there was the last leave-taking, and Moffat started for his distant home."

Lobengula succeeded his father Umziligazi ; the progress and end of his evil reign are fresh in the mind of everyone.

As soon as the Ngoni had crossed the Zambezi it is said they were in the country of the Senga. These are not the Senga now living on the Loangwa further north, of whom more hereafter. Their languages are quite distinct. The Senga

Ngoni Warriors.

tribe being an easy prey to the Ngoni (who must now have been very numerous, composed of the original stock, and the Swazi, Tonga, and Karanga additions by the way) at once submitted and were incorporated. They rested in this district, eating up the food of the country and initiating the Senga into the use of their weapons of war, the shield and spear.

Leaving the country of the Senga, considerably increased by the addition of that people, they journeyed north, evidently along the watershed of the Loangwa river, until they came into the district named Matshulu which was inhabited by Tumbuka, who went under the name of Amamatshulu. The Tumbuka tribe had evidently covered a wide area, but as they lived in small villages of two or three huts they may not have been so very numerous. The Tumbuka are a very industrious agricultural people, and having been unable to resist the Ngoni horde they submitted, and laboured to supply the needs of their conquerors. The Ngoni are said to have lived for a comparatively long period in the Matshulu district, and here began a condition of things in Zongandaba's following which may have delayed their northward progress for three or ten years, as it is variously estimated by natives. It was at any rate a " killing time,"

which has impressed itself on the minds of the
people to this day. Zongandaba had no doubt
conceived that he could best conserve the interests and combine the influence of those he
had conquered and incorporated, by appointing
certain of each tribe as his advisers. He had a
council composed of Tonga principally, and his
original followers began to be jealous of them,
and of Zongandaba's evident love for them. The
Tumbuka, adepts at witchcraft practices, they
impressed into their service. Charges of witchcraft were brought against the leading members
of the Tonga tribe, and by the aid of the Tumbuka doctors and their incantations, Zongandaba
was incited repeatedly to organize an army and
destroy a whole village at a time. None in the
village were spared, and during their stay in
Matshulu nearly all the Tonga were massacred
in this way. To this day to say "People were
killed at Matshulu" is to emphasise a large
number as quoted. It was evident that discontent and thirst for power had appeared to
disorganize the hitherto united band, and it is
said that, after this, Zongandaba became very
despotic and approached to having the character
of Chaka. Such a heterogeneous collection of
men would doubtless produce a despotic ruler.
Only one or two Tonga who had left their own

country were spared to the end of the Ngoni wanderings, but some of their children are still living.

Having again taken their road northward they came to the district they name Mapupo, inhabited by the Sukuma. The district lies near the south end of Tanganyika and is now on the maps as the Fipa district. Here Zongandaba died, after which the tribe suffered several disruptions. While in this district, and combined, they carried war northward on the east of Tanganyika; eastward as far as the Nkonde tribes at the north end of Lake Nyasa, and south-eastward to the Henga, then living in the mountainous country near the Rukuru river, a few days' journey from their present location, which was the country of the Tumbuka originally.

At the disruption the chief sections were: 1. That under Ntabeni which went northward on the west side of Tanganyika, where in 1879 they were heard of by the late Mr Stewart. 2. Ntutu led another section northward on the east side of Tanganyika; of these Stanley in his "Through the Dark Continent" says, "No traveller has yet become acquainted with a wilder race in Equatorial Africa than that of the Mafitte (Maviti) or Watuta. They are the only true African Bedawi; and surely some African Ishmael must

have fathered them, for their hands are against
every man, and every man's hand appears to be
raised against them. . . . The Watuta became
separated from the Mafitte (Maviti or Ngoni) by
an advance in search of plunder and cattle."
They carried war and bloodshed over a vast
extent of country, as may be seen by a glance
at a map of Central Africa. Considering that
they were only a sub-section of the Ngoni, the
following graphic description of their expedition will indicate the tremendous wave which
Chaka set rolling over twenty-six degrees of
latitude. Mr Stanley continues: "The separation (*i.e.* of the section referred to above as
led by Ntutu) occurred some thirty years ago
(1840). On their incursion they encountered
the Warori who possessed countless herds of
cattle. They fought with them for two months
at one place, and three months at another; and
at last, perceiving that the Warori were too
strong for them—many of them having been
killed in the war, and a large number of them
(now known as the Wahehe, and settled near
Ugogo) having been cut off from the main body,
—the Watuta skirted Urori, and advanced north-
west through Ukonongo and Kawendi to Ujiji.
It is in the memory of the oldest Arab residents
at Ujiji how the Watuta suddenly appeared and

drove them and the Wajiji to take refuge upon Bangwe Island.

" Not glutted with conquest by their triumph at Ujiji, they attacked Urundi; but here they met different foes altogether from the negroes of the south. They next invaded Uhha, but the races which occupy the intra-lake regions had competent and worthy champions in the Wahha. Baffled at Uhha and Urundi, they fought their devastating path across Uvinza and entered Unyamwezi, penetrated Uzumbwa, Utambara, Urangwa, Uyofu, and so through Uzinja to the Victoria Nyanza, where they rested some years after their daring exploit. They ultimately returned and settled in Ugomba, between Uhha and Unyamwezi. They are called by the Nyamwezi Ngoni."

3. The third section is that over which Mombera was appointed chief. Mtwaro should have been chief, but he resigned in favour of Mombera, as being of a quiet disposition; he felt the burden of ruling such a jealous, discontented people as they had become would be too great for him. Under Mombera there were his brothers Mtwaro, Mperembe, Mpezeni and Maurau. This section moved eastward to a place called Tshidhlodhlo, the locality only being known now as somewhere about the north end of Nyasa. Here a great

battle was fought, and the Gwangwara, overcoming the others under Mombera, drove them back in a southerly direction. The Gwangwara, in settling on the east side of Lake Nyasa, form the outmost ripple of the wave on that side, and they have carried fire and sword southward into Yaoland, and as far as Masasi, the station of the Universities' Mission. Those under Mombera at this point suffered a disruption. Mperembe and Mpezeni broke off. Mperembe returned to attack the Bemba to the south of Tanganyika, and Mpezeni went south and settled where he now is, west of the southern extremity of Lake Nyasa. Chiwere, a head man, went off with a following, and settled west of Kotakota. Mombera's division first settled in Henga (the lower reaches of the Rukuru river), and subjugated the Henga section of the Tumbuka tribe, ultimately entering the Tumbuka country proper, on the south-west of Choma mountain. Being joined again by Mperembe, they have continued to occupy the valleys of the Lunyangwa, Kasitu and Rukuru. They defeated and began to govern the Tumbuka and Tonga on their arriving there, and have only a few years ago given up their predatory habits.

What might not have happened had the dawn of this century witnessed the enthusiasm of the Christian Church in the cause of foreign missions

LUNYANGWA RIVER, EKWENDENI.

which is a feature of its close ! What achievements for Christ there might have been ! Here we stand at the Zambezi and look back at the reigns of Dingiswayo, Chaka and Zwide, and see the rise and fall of kingdoms ; rivers of blood shed; a million or more massacred, condemned to cannibalism, or to death by starvation ; fathers slaying their children, and children their fathers ; and God's fair earth made worse than hell—all for want of the Gospel. We see before us a horde of barbarians, their faces set to the north, who, over hundreds and hundreds of miles, are to spread death and desolation ere the Gospel comes to them to make them new men. Had the Gospel been brought to Dingiswayo's kraal then, what a different history of South and Central Africa could have been written ! There was then a more open door to these regions than there has been in these later days, according to the history of missions in Zululand, Matabeleland, Gazaland, the Upper Zambezi, Nyasaland, and away round Nyasa by the country dominated by the Gwangwara, who are Ngoni, down through Yaoland, for all were affected by the convulsions induced in Chaka's time. We read of Dingiswayo in the beginning of this century opening a trade with the Portuguese at Delagoa Bay, giving liberal rewards to his people for inventive or imitative genius displayed in the production of things with which he might trade with the Portuguese, and having a karosse manufactory, in which a hundred men were employed. These were days of peace and industry such as have not been found anywhere else on the arrival of missionaries in those regions. Again, in the days of Chaka, over whom one or two Europeans (Messrs Fynn, Farewell, and Isaacs) seem to have had great influence, gained by fair dealing and medical skill, one of them wrote : "On one occasion, as I have before related, when we communicated to him our opinions on the existence of God, who made the world, and of a future state, and told them that by a knowledge of letters all our confidence of being immortal beings had arisen, he expressed surprise, and wished much that the doctors or missionaries would come to him, and teach him to acquire this knowledge. The greatest state of ignorance on this sublime subject pervaded him. But I have ever been impressed forcibly from the desire he manifested to have among his people missionaries whom, he said, he would protect and reward, that he might have been brought to some sense of reason on this important point, so necessary for the promoting of civilization." But the Church of Christ was at the time ignorant of her duty, and was not impressed by the opportunity of extending her Lord's kingdom.

CHAPTER II

THE NATIVES AND THEIR COUNTRY

THE physical features of Ngoniland may be denoted in a few words. Situated about 4000 feet above the sea-level it has little or nothing to suggest its being in the tropics, save the daily course of the sun and the periodic rains. There are no broad sluggish rivers whose muddy banks are covered with mangrove thicket, above which rise giant trees and stately palms such as are usually associated with pictures of tropical scenery. Leaving Lake Nyasa at an altitude of 1500 feet we have to cross the broken mountain ranges, rising in some cases to 7000 feet, which form the eastern boundary of Ngoniland. From the heights we behold hundreds of square miles of open undulating country, whose low wooded hills run north and south for most part, the broad valleys being traversed by streams which become roaring torrents during the brief rainy season, but at other times are small and easily forded. Looking over the

country at our feet, we are struck by its treeless-
ness, save on the crowns of the low hills. Here
and there we find single large trees and, at
intervals, dark green patches which look like
fields of green corn, but which are in reality
patches of bush composed of fresh shoots from
the roots of trees cut down, which features
denote dry unfruitful soil not worth tilling. It
is evident that, at one time, the whole country
was covered by dense forests of large trees, which
have been ruthlessly cut down for fire-wood, or,
as is more frequently the case, to be burned on
the ground as manure for new gardens. The
intervening ground, if viewed in the dry season,
appears as bare, whitish, or yellowish-red soil,
as the extensive gardens are then empty and the
grass burned up. It is not easy to pick out the
villages as the colour of the dried thatch accords
with that of the bare ground and renders them
not readily visible. The most conspicuous
feature of the district is the innumerable ant-
hills scattered over the plains. Seen from a
distance they resemble stacks of hay in a field.
They are the product of the white ant, the most
destructive pest we have, a full account of which
is given in a most interesting way in Prof.
Drummond's "Tropical Africa." The ant-hills
in Ngoniland are larger than any to be seen

elsewhere. They are not the turret-shaped variety to be seen in the low countries, but are huge mounds in many instances 50 feet in circumference at the base and 20 feet in height. The clay composing these mounds is very suitable for brick-making, and from even one ant-hill a whole Mission station could be built.

The villages are situated near the streams or fountains. The native has no idea of bringing water to his town save by the usual beast of burden—woman, and so the presence of water decides where the village is to be built. He can drive his cattle far enough to pasture, or go miles and cultivate his garden, but water which is needed every day has to be carried, and the women who have to do that have some voice in the choice of a site for a town. The low hills form natural divisions between chiefs' and sub-chiefs' districts,. and consequently, while Ngoniland is perhaps 100 miles long by 60-80 broad, the villages are mainly in groups around the large town of the chief or sub-chief, and are easily overtaken by district schools and evangelistic agencies.

The towns and villages are not permanent locations. Every three or four years the inmates find it necessary to make new homes, and a fresh start in life as regards domiciliary comforts.

The white ant attacks the wood and grass of the hut; the bugs, tampans and jiggers, disturb the peace of the inmate; and the accumulations of filth around the village make life unbearable even to the native; he is forced to seek a new home.

Removing a village to a new site was one of the great events in the history of the people. It marked a division in his calendar and became a point by which he could locate events. It was one of the occasions when he had to be religious, and so the removal was inaugurated by certain religious rites. The cattle are the sustenance and the bond of the family, the village, and the tribe. The care of the cattle in the new town was first seen to. The size of the fold having been decided upon, and marked off by making a circle, it was built of trees and shrubs, at first of a temporary nature, because by tradition it had to be begun after sunrise, finished, and the cattle folded before sunset, on the same day. When the cattle were driven in, the religious ceremonies conducted by the divining doctor were further developed, by selecting a certain beast as a sacrifice to the village ancestral spirit. This beast would ultimately be killed for the spirit, and eaten by the people when the village was occupied. Although many religious rites of the people appear to us grotesque and unreal, yet a close examination of them proves the existence of their belief in a Providence, a Judge, and an Almighty King, but we cannot stop to unfold the matter here. The huts of the people are built in circles around the cattle-fold. Like everything the native makes they are circular, and he points to the sun, moon, and horizon as a reason why they should be so. A few sticks set in the ground and plastered inside, with a wattled roof covered with grass, constitutes the native hut. He does not use it as a shelter from the sun but from cold, and its circular form reflects heat and renders it comfortable in the cold nights which are experienced on the hills.

The size of the hut depends upon the position of the master; it is from 10 to 20 feet in diameter, but the walls are not more than from 4 to 8 feet in height. The roof comes down nearly to the ground, and so a cool verandah is formed, under which the inmates can enjoy their siesta, or congregate on wet days to indulge in their favourite pastime—gossip—or perform their toilet, the women requiring a long time, owing to their manner of dressing the hair. The huts are single-roomed of course, the inner part being the storehouse for seed, corn, pots, and

other utensils required in the daily round.   The
fire is made in a circular depression in the middle
of the floor, and the cooking-pot is set on three
stones above the fire, which is always of wood.
The smoke finds an exit by the door or through
the roof, and the rafters are covered by soot
which protects them from the attacks of white
ants.   One can tell the direction of the prevailing
wind, by the colour of the outside thatch being
browned by smoke on the leeward side.   In days
by-gone the floors of the huts of the better
classes were like polished ebony.   Clay was
beaten hard and smooth while drying, and after
being polished by rubbing with smooth stones,
the floor was smeared with ox-blood and polished
again.   In ordinary cases the floors and open
space in front of the hut were smeared with
fresh cow-dung subsequently scraped off by hand;
this left a clean and cool floor free from dust
in which fleas could breed.   The brick floors of
many Mission houses are regularly treated in the
same way, and it is found to be a good plan for
preserving the floors intact.   In the days when
every Ngoni was a warrior, it was the work of
the women to build and repair the huts, as well as
cultivate the gardens, but now the men share the
work, and all that the women do is to collect grass
for thatch, plaster the walls, and make the floors.

But before the huts are built—as the village is always built in autumn—the grain-stores have to be erected for the crops to be reaped. They are made by plaiting reeds into huge baskets 5 or 6 feet high and as many in diameter, which are placed on platforms a foot or more from the ground. Sometimes they too are plastered, but only on the outside, and when the mealies or millet stored in them have been well dried, a grass roof is put on prior to the rains. These grain-stores are built between the huts and the cattle-fold. The huts are arranged in groups walled off from each other by reed fences, so that each man with his wives' huts, and those of his slaves, if he has any, has a distinct locality in the village. The huts of the headman or chief and his seraglio and slaves, are situated always at the opposite side from the cattle-fold gate, from which a broad road leads to the watering or pasture. The space at the kraal gate is the public room of the village where anyone may go, and where we usually have our services, but inside the cattle-fold all *indabas* (cases) are talked, and the village dances take place.

Such is the description of a native village. Around the huts the smooth beaten ground is swept every day, and when once inside the village, one's sensitiveness is not offended, but the serious matter is the approach. Good for the natives is it that their bodies cannot always endure the incessant attack of certain insects inhabiting the huts, and that they are compelled every three or four years to build a new village and burn everything connected with the old one. There is not the slightest attempt at sanitary arrangements. The ashes from the fires, the refuse of maize, the sweepings of the village, and filth of all kinds find their place just round the village behind the outer row of huts. The state of filth around is indescribable. After a year or two the *tampan*, one of the greatest and most prevalent pests of Africa, multiplies in the huts, and so at length, more from that than because of the general collapse of the village, the natives have to make a new one. The tampan is a thousand times more annoying than the bug of which also there is usually a good supply. It is larger when full grown than a sheep-tick, of a dirty-grey colour, and so tough as not to be easily killed by crushing. The sight of them, even before one has experienced their bite, is most repulsive. They are not to be seen during the day as they enter the cracks in the roughly-plastered huts, or hide in the roof, but no sooner has one lain down, than they come out and feed off him. Their bite is very irritating, and has the reputation of producing fever, dysentery, and other troubles. The effect of the bite appears to be dependent on the physical condition of the individual at the time of the attack. I have been bitten when there have been no effects perceptible except the discomfort locally. At other times a night or two in a native hut has almost completely laid me down—the feeling of malaise and tendency to sickness were very pronounced. The tampan seems to be common all over Africa, and a species from Egypt is named *Argas savignyi*, with which those in Central Africa are closely allied. The sleeping-place of native servants on the stations cannot be kept free from them. The boys bring them from the villages in their clothes, but ordinary care prevents their entrance into the missionary's rooms. Indeed from that and other commoner organisms, whenever I returned home from a tour on which I had to reside in native huts, I was put in quarantine as a precaution.

When the natives leave their old village the huts are burned down, except those belonging to deceased persons, which are left to fall to pieces, as the spirits are supposed still to visit them. On the site of an old village for many years they sow maize, and I have seen it 12 feet high and growing so closely together as to be scarcely penetrable.

Let us spend a day in such a village. The native is an early riser. Ere the sun has appeared, men and women are out of doors. The cow-herds have gone to milk the cattle before driving them out to the bush, where they browse all day and are brought home at sunset, when they are again milked. The women set off to the river with a big earthen pot on the head, and return with it full of water—such-like exercise giving the native women that grace of carriage which would be the envy of ladies in civilized countries. The native woman can carry twice as much as a man on her head. If the village is dependent on water from a fountain it is "first come best served." I have been marching through a fountain country at four o'clock in the morning, and seen women and girls running to the fountains at that hour, in hope of finding sufficient water before the others come. Then the woman has the firewood to gather, the maize to pound in a wooden mortar and grind into flour for the evening meal. She has to find the *umbido* (green herbs) which, in the absence of meat, is required as a relish with the stiff maize porridge which is the staple diet of man, woman, and child. She has a large part of the day in the dry season in which she may gossip with her neighbours, or lie down and sleep in the cool verandah of her hut. As evening comes on she has again to visit the river with her water-pot, and cook the food for the men, who eat apart, no woman venturing to eat along with her husband or in the presence of a man. In the rainy season she has hard work indeed, having to work in the gardens in addition to her household duties. The one thing a woman tries to excel in, and gain a reputation for, is the making of beer. Brewing is solely woman's work. She is privileged to preside at the beer-drinking, and usually ends all by becoming intoxicated. She may not eat with her husband or his friends, but she may get drunk along with them. At other times she has to reverence her husband by not pronouncing his name, unless she swears by it, but at beer-drinkings no rule binds her save that her beer ought to make those who partake of it drunk. These beer-feasts end in quarrels and evil of every kind.

A very bad custom obtains in connection with planting and reaping which produces much drunkenness. The meagre hoeing given to the ground necessitates the cultivation of vast stretches of garden ground, in order to plant the year's supply of food. To get the ground hoed and planted, householders, who have many gardens, invite labourers by carrying large pots of strong beer to the garden. There is no lack of willing workers who drink and shout and, in the end, quarrel and fight, sometimes laying each other's heads open with a blow from their hoes. These scenes are utterly degrading and nothing but a heartier desire for honest work by each owner of a garden, and thoroughly cultivating a smaller tract, will put down these scenes. Then when hundreds of baskets of maize and other grain have to be carried home, or the beer crop cut stalk by stalk and gathered, help is again required, and a beer-drinking brings together the workers. Our teachers have set their faces against this vile custom and have instituted a feast—mutton or goat-flesh and porridge—when help is required, and thus a step towards a better state of things has been taken.

The work is done principally by the inferior wives, if a man has more than one. The head-wife, however, is the overseer and, in a polygamous household, if her favourites are not for the time being also her husband's favourites, she makes it hot enough for those whom she considers to be too attractive to him. There are frequent brawls, but should a man strike a wife or any woman he is branded indelibly as a bad man and may as well go and hang himself. The multitude of his wives do not bring him peace.

The wordy warfare is often sharp and long and, in a measure, he has to guard his words lest a wife be driven away to her father's house, in which case, if the cause was sufficient, she may remain away having as her portion the cattle that were paid for her when she was betrothed. I have seen a man hurrying after a raging wife who was *en route* for her father's house, and it was anything but a dignified position even for a native to be in. On one occasion a man came to beg cloth from me to settle an *indaba* he had. On enquiring I was told that one of his wives had been offended at some scolding he gave her and had gone to her former home. She had now repented and was willing to return to her husband, but her father's people would not allow her unless he first paid something for having caused her to run away. I enquired how many wives were left to him and he said he had still five. I advised him to let the run-away one stay where she was, but the great matter for him was that she represented so many head of cattle and he could not lose them as, by having children by her he could give them out in marriage and so get his cattle-fold restocked. There was no room for the sentiment of love. It was purely a mercantile transaction. Here is a native's description of a household squabble :—

"This is a story about wives. A man had five wives and they were quarrelling among themselves. One said to another, 'You are all right since our husband loves you only. As for us he does not love us at all.' So they seized each other and fought, one of them being greatly hurt in the quarrel over their husband. The husband said, 'I love you all, my wives.' One replied, saying, 'You just love one of your wives.' Others said, 'What did he take us from our father's house for, seeing he only loves one?' There was war very often."

When evening comes the principal meal of the day is eaten. It consists of maize flour made into a very stiff and very partially cooked porridge, which is accompanied by a relish composed of meat with a little salt, green vegetables or dried herbs. What bread is to us this porridge is to the native. It matters not however freely he eat, for instance, of flesh and vegetables, he will complain of hunger unless he has had his quantity of porridge. At meals the women and girls eat by themselves in one part of the family compound or open space, and the men who are usually to be found in the cattlefold may have theirs along with the boys there. When the meal is over there is not much labour clearing the table or in the scullery afterwards.

The porridge has been cooked in one huge pot, and the portion for the women put into a broad flat dish, with the relish in a small earthenware pot, and that for the men and boys has been served up in the same way. They all sit round and, dipping the fingers in the heap of porridge, take a little which they roll into a ball, dip it in the relish and literally pitch it into the mouth. They do not chew it, and hence the manifold digestive disturbances the natives are liable to. The delicacies of civilisation are said to have made men more unhappy and unhealthy than is the simple untutored savage. My experience is that civilised people have not so much sickness as natives. Their splendid ivories are made much of, but, as I have seen a few hundred mouths, the front teeth are usually the only ones preserved.

When the evening meal is over, if it is the dry season and a moon present, the youths and maidens of the village go to the cattle-fold to the dance, which is a recreation much liked by the natives. The Ngoni, unlike the Tonga and Tumbuka peoples, have no obscene dances, and on a clear evening, when all around is still, it is very enjoyable to listen to their song accompaniment (from a distance). It is then that the glamour of native life is thrown over the casual visitor, and perhaps it is excusable that he goes away filled with the idea that the native spends an idyllic life, has no care, and is always happy and free. True, there is apparent peace and joy in the village as the young people, not infrequently joined by many of the mothers with babies on their backs, join in song and dance for an hour or two after sunset. But it is only one phase of native life, which does not, to those who are behind the scenes, cover the unhappiness, the slavish fear of evil spirits, the often cruel bonds of heathen customs, and above all the secret immorality, lying, stealing, and often murder, which abound in every native community.

The song is the principal thing—not the dance. The dance is the accompaniment of the song, and not *vice versâ*. Their songs are well-nigh unintelligible to a stranger, as they consist of short statements relating to some incident in the everyday life or history of the people, and without a knowledge of those incidents one cannot understand them. From them, however, one may obtain a very minute record of the people's history. The men, with dancing-sticks in their hands, held erect, form one line, and the women form a line some distance apart from, and opposite to, the men. All sing heartily, and the dance consists in merely striking the ground with the feet, while the sticks are waved overhead, with certain movements of the body and head carried out in unison, the whole combined forming a not unpleasing, although unrefined exhibition. The song, as heard from a distance, is not without artistic effect as the high-pitched voices of the women, usually very musical, and the deeper voices of the men rise and fall in the evening stillness in musical cadence. In some of the songs there are dialogues, the men and women speaking to each other in rhythmical notes. In these dialogues the music is not unsuited to the subject. In some songs the maidens take up, it may be, a taunt against the young men concerning some war exploit, domestic *fracas*, or playfully assert that the young men of their village are inferior to those of some other village. To this taunt in song the young men reply in notes suited to their indignation at the charge. Thus the song goes on, while the rhythmic gestures and beating of the ground with the feet add zest to the subject. At certain stages in the song the words are dropped, and the women continue the tune in a low, humming voice, while the movements of the men are continued; and then, at another stage, the women clap hands in unison, but always in two parts, with a slight interval of time, so that the sound is doubled and accentuated. The dance forms a suitable occasion for the youths of the village to show themselves off in front of the young women, whose favour they may be anxious to obtain.

Such is the village dance; but in the dry season, after the crops have been reaped, there is a kind of competitive dance engaged in between two villages. Without warning, the young people of one village will come to another village, dressed in all their best things. They enter the cattle-fold singing, and begin to dance. Those of the village visited who are within call are quickly summoned to engage the strangers, and they are prepared to begin to dance when the other party stops to rest, the desire being to out-dance the other by holding the field as long as they can, as well as to have the best singing and most perfect movements. Thus they go on, one party after the other, during the whole day, and when the sun has well declined, the strangers return home, singing gaily all the way.

The daily life of the men is soon described. They have usually no work to do. Their day is spent in talking, taking snuff, and drinking beer. They may do a little hoeing in the busy season, and cut the trees where a new garden is being made, but that is about all. The introduction of labour by the Mission has effected a great change, as the men who were wont to go out raiding other tribes during the whole of the dry season, are now found eager to obtain work. Some few are found who, of their own free will, devise work, and are always busy since there are trades found among them.

The men's place is the cattle-fold, where they spend their day, and a stranger visiting a village goes to the gate to await the salutations of the people, and to be enquired of as to his business before he is conducted to the house of the party he may have come to see. There is a well-defined etiquette observed throughout the community. It is a great offence for one to sit down opposite the door of a hut. A native's house, as well as a Britisher's, is his castle, and no one dare enter uninvited. Neither may one sit down near the house without giving warning by a cough, an exclamation, or by salutation, as eaves-dropping is a crime which is abhorred by the natives.

One of the pretty sights about a native village in the evening is the folding of the cattle. As the sun sinks the cattle begin to turn homeward. The boys who tend them have reeds which they cause to emit a not unmusical sound—the different cattle-herds having differently pitched reeds —by manipulating the open end as they blow through, and all sounded together produce a simple, sweet music. The cattle collect together where they have been grazing as the boys blow their reeds, and wend their way home for the evening milking and to rest over night in the open fold. The old Ngoni were wholly a pastoral people, and only in recent years have gone in for agriculture to the extent they now do. Before the cattle plague the herds were numerous and large, but now there are only tens where before there were hundreds. The cessation of war raids also accounts to some extent for the decrease in the number of cattle owned, as cattle-lifting was a constant occupation in the dry season.

CHAPTER III

IT is a mistake to suppose that even among barbarous tribes, such as the Ngoni, all their customs are bad. There were, before Christian teaching began to influence them, many things which were admirable. Those traits of character and customs so readily seen by strangers, the observation of which has so often led travellers to believe that the state of the untutored savage was happy, free and good, are nevertheless found alongside lower ways of living, and a grossly immoral character, which are not only the obstacles to Mission work but its *raison d'être*. It is not our purpose, meantime, to state or explain fully the customs of the people, all of which have an interest from the anthropological point of view, but to present a brief sketch of those which stood out as hindrances to the progress of our work, and which, being bad, had to succumb to the influences of the moral and spiritual teaching of the gospel. There are many customs so grossly obscene that we cannot enter upon a statement of them. I avail myself of a letter from my colleague, Rev. Donald Fraser, which he recently sent home, describing what he witnessed in an out-lying district of Ngoniland in connection with the initiation customs at the coming of age of young women.

" Leaving these bright scenes behind, I moved on west into Tumbuka country to open up new territory. But scarcely had I turned my back on Hora when I began to feel the awful oppression of dominant heathenism. For a few days I stopped at the head chief's village, where we have recently opened a school. The chief was holding high days of bacchanalian revelry. He and his brother and many others were very drunk when I arrived, and continued in the same condition till I left. Day after day the sound of drunken song went up from the village. Several times a day they came to visit me and to talk : but their presence was only a pest, for they begged persistently for everything they saw, from my boots to my tent and bed. The poor, young chief has quickly learned all the royal vices—beer-drinking, hemp-smoking, numerous wives, incessant begging. I greatly dread lest we have come too late, but God's grace can transform him yet.

" When we left Mbalekelwa's we marched for two days towards the west, keeping to the valley of a little river. Along the route, especially during the second day, we passed through an almost unbroken line of Tumbuka villages. At every resting-point the people came to press on us to send them teachers, and frequently accompanied their requests with presents. When at last we arrived at Chinde's head village, we received a very cordial welcome. Chinde (a son of Mombera) did everything he could to convince us of his unbounded pleasure in our visit. For three or four days we stayed there, and were overwhelmed with presents of sheep and goats, and with eager requests for teachers. Leaving this hospitable quarter, we had a long, weary march through a waterless forest, in which we saw the fresh spoor of many buffaloes and other large game, and heard a lion roaring in front. Late in the afternoon we reached Chinombo's and remained for other three days. Here again, we were well received and loaded with presents.

" This whole country to the west is still untouched. That the people are eager to learn is evident from their urgent requests. That they sadly lack God, and are living in a dreadful degradation, became daily more and more patent. I cannot yet write as an inner observer. Tshitumbuka, the language spoken there, I am only now beginning to learn. Yet the outer exhibitions of vice and drunkenness and superstition were only too painfully evident.

"Often have I heard Dr Elmslie speak of the awful customs of the Tumbuka, but the actual sight of some of these gave a shock and horror that will not leave one. The atmosphere seems charged with vice. It is the only theme that runs through songs, and games, and dances. Here surely is the very seat of Satan.

"It is the gloaming. You hear the ringing laughter of little children who are playing before their mothers. They are such little tots you want to smile with them, and you draw near; but you quickly turn aside, shivering with horror. These little girls are making a game of obscenity, and their mothers are laughing.

"The moon has risen. The sound of boys and girls singing in chorus, and the clapping of hands, tell of village sport. You turn out to the village square to see the lads and girls at play. They are dancing; but every act is awful in its shamelessness, and an old grandmother, bent and withered, has entered the circle to incite the boys and girls to more loathsome dancing. You go back to your tent bowed with an awful shame, to hide yourself. But from that village, and that other, the same choruses are rising, and you know that under the clear moon God is seeing wickedness that cannot be named, and there is no blush in those who practise it.

"Next morning the village is gathered together to see your carriers at worship, and to hear the news of the white stranger. You improve the occasion, and stand, ashamed to speak of what you saw. The same boys and girls are there, the same old grandmothers. But clear eyes look up, and there is no look of shame anywhere. It is hard to speak of such things, but you alone are ashamed that day; and when you are gone, the same horror is practised under the same clear moon.

"No; I cannot yet speak of the bitterness of heathenism, only of its horror. True, there were hags there who were only middle-aged women, and there were men bowed, scared, dull-eyed, with furrowed faces. But when these speak or sing or dance, there seems to be no alloy in their merriment. The children are happy as only children can be. They laugh and sing, and show bright eyes and shining teeth all day long. But what of that? Made in God's image, to be His pure dwelling-place, they have become the dens of foul devils; made to be sons of God, they have become the devotees of passion.

"I have passed through the valleys of two

little rivers only, and seen there something of
the external life of those who can be the children
of God. The horror of it is with me day and
night. And on every side it is the same. In
hidden valleys where we have never been, in
villages quite near to this station, the drum is
beating and proclaiming shame under God's face.
And we cannot rest. But what are we two
among so many? O men and women, who have
sisters and mothers and little brothers whose daily
presence is for you an echo of the purity of God,
why do you leave us a little company, and
grudge those gifts that help to tell mothers and
daughters and sons that impurity is for hell, and
holiness alone for us !

" ' How long, O Lord ! how long ? '

"I send you this account of a missionary
journey. Would that my pen could write the
fire that is in my soul ! It is an awful thing to
sit looking at sin triumphant, and be unable to
do anything to check it. Calls for teachers are
coming from every side, but we cannot listen
to them at present—our hands are more than
full."

The letter refers to the custom as it obtained
among the Tumbuka and Tonga slaves, and it
presents an awful picture of moral degeneracy

which was all too commonly seen on such occasions all over Central Africa. Although the Ngoni practice was less openly obscene yet the occasion was one of unspeakable evil, extending over several days, on which both sexes were accorded full licence for every unholy passion.

In like manner in connection with marriages —especially of widows—and the birth of twins; when armies returned from war and the purification ceremonies took place, practices which are not meet to be described were unblushingly engaged in. What in Christian lands is held sacred in heathen lands is too often the common property of young and old, and where public opinion is devoid of the moral sense we cannot look for elevation from within.

One of the greatest social and moral evils among the tribe is polygamy. The evils are seen among all classes, for as the tribe existed by raiding other tribes, all who could bear arms might possess themselves of captive wives. Among the upper classes the rich held the power to secure all the marriageable girls in the tribe, by purchasing them from their parents for so many cattle. The practice of paying cattle was not in all cases wholly bad, but the tendency was to outrage the higher motives and feelings, especially in the women who often were bargained for by their parents long before they entered their teens. The cattle paid to the father of the bride formed a portion which she could claim and have as a possession, in the event of her being driven away by the cruelty of her husband, and, in the absence of a nobler sentiment, it was in some degree a safeguard of the interests of the wife. But upon no grounds, social or moral, could such a practice be defended. It is inimical to the true morality of marriage, and consequently to the progress of the race. It is no uncommon thing to find grey-headed old men, with half-a-score of wives already, choosing, bidding for, and securing, without the woman's consent, the young girls of the tribe. Disparity of age, emotions and associations, make such unions anything but happy, and nowhere do quarrels and witchcraft practices foment more surely than in a polygamous household. A man's wives are not all located in one village. He may have several villages, and from neglect young wives are subject to many grievances and temptations, so that it is no wonder they age in appearance so rapidly. They are often mal-treated by the senior wives, who, jealous of them, bring charges against them, and, in the hour when they should have the joy of ex-pectant motherhood, they are cast aside under some foul charge, without human aid or sympathetic care. On more than one occasion I have been called by a weeping mother to give aid to her daughter in such circumstances, when, if a fatal issue resulted, she and her family would have been taken into slavery and their possessions confiscated. Only those who spend years among them and are their trusted friends can tell of that and countless other unholy and inhuman things, which result from the custom of polygamy as it exists.

Flippant writers on such customs, especially some travellers who had not the opportunity of becoming acquainted with the people, state that polygamy is, in the savage state where there is an absence of higher motives, a safeguard of morality. It is, however, far from being so. Men with several wives, and many of the wives of polygamists, have assignations with members of other families. I have been told by serious old men that such is the state of family life in the villages that any man could raise a case against his neighbour at any time, and that is one reason why friendliness appears so marked among them—each has to bow to the other in fear of offending him and leading to revelations which would rob him of his all.

The belief in witchcraft is the most powerful

of all the forces at work among the tribes. It is a slavery from which there has been found no release. It pervades and influences every human relationship, and acts as a complete barrier to all advancement wherever it is found to operate. No matter whether it be master or slave, chief or subject, parent or child, he has to bear this yoke which may at any moment crush him. He lives in fear. If he is sick it is not a question of how he may be cured, but of who has bewitched him ; or if his plans are frustrated what evil spirit has been moved against him. The reason for his apparent laziness is the fear that, if he become possessed of goods, his circumstances will excite jealousy and bring on him accusations of witchcraft, and death as a result. It is productive of unrest, cruel treatment, and great loss of property and life.

The *itshanusi* or witch-doctor lives upon the credulity and slavish fear of the people. He is either self-deceived or a base impostor, but his power for evil in a tribe is unlimited. He is reverenced by all classes, and although one may hear whispers of a want of faith in him and his incantations, no one would dare to oppose him in public. Wicked men and chiefs make use of him and his immunity from punishment to " remove " any person who is disliked or whose possessions have rendered him opprobrious to them, and a chief or headman's unjust demands may be bolstered up by an appeal to his easily-bought action. They aid despotic chiefs in governing a discontented people, and from the deep religious feeling which the people have in regard to the presence and power of the ancestral spirits with whom the *itshanusi* is believed to be in communication, they are ready to acknowledge even that which may be to their hurt.

As to their belief in witchcraft I might refer to what I have observed in the course of my practice of medicine among the people. No sooner is it concluded that a person who is sick has been bewitched, than the friends around talk of it without constraint in the presence of the patient. Sometimes they may carry him about from place to place in the hope of cheating the charmer, but the effect on the patient is very marked. He seems to conclude that he is to die, and he evinces no fear or anxiety in view of death. He assumes an unnatural stolidity, despair, and what might be termed resignation. Although his imminent death is talked of freely before him he has no fear or complaint. He shows no desire to fight for life, but with an inhuman want of hope or desire for recovery he awaits the end. The thought that he is bewitched seems to deprive him of all natural clinging to life. Even among the youthful of both sexes there is that want of hope, when once the elder people have declared they have been bewitched.

In connection with charges of witchcraft, the poison ordeal is the final and too often calamitous sequel. Before the light of Christian truth came to them, and has, even where the doctrines are not wholly embraced, done away with this great evil, the number annually killed by drinking the *muave* cup cannot be estimated. Anything a man possesses, about which there is any mystery, may give rise to a charge of witchcraft. If a man is found walking near a village at night he is charged with evil intentions. If one possesses himself of an owl or other night bird or animal, he is supposed to work evil by means of such, and is charged forthwith. When sickness or death comes into a house or village someone is blamed. The *itshanusi* is called, and there are not wanting those who in their talk reveal in what direction the thoughts of the people lie, and so he names someone, which decision at once appears reasonable to the people and is accepted. Often the witch-doctor has emissaries secretly employed to find out what he wants, and, acting upon information thus obtained, he appears to the people to be acting upon communications he has received supernaturally. Sometimes he does more to influence their imagination and make themselves name someone than by himself doing so directly. I have known several witch-doctors, and have come to regard them as shrewd individuals, certainly more given to thought than the community generally, and who traded on the superstitious fears of the people, who seldom exercised their reason in connection with ordinary occurrences. On many occasions men and women have sought refuge at the Mission station when accused of witchcraft and under sentence of death. On one occasion, during a trial which took place at a village near the station, when the *itshanusi* was performing his incantations and condemned a man, he broke away from the crowd and ran towards the house. He was followed by a crowd of men and boys clamouring for his life, and being overtaken, was clubbed to death before our eyes ; his body was ignominiously dragged back to the scene of trial, where it was subjected to gross indignities.

On all occasions of administering the poison cup we tried to stop it. Sometimes we were successful and sometimes we were not. Sometimes we were able to prevail upon them to substitute dogs or fowls for the human subjects, and then it was possible for us to watch the proceedings. These were occasions on which the whole community turned out. The friends of the accused were very few on such occasions, and the people jeered the unhappy wretch and engaged in song and dance while he had to stand alone and prove his innocence by vomiting the poison, or, by death from the poison, confirm the truth of the charge against him. When the poison began to take effect, as seen in the quivering and collapse of the culprit, it was the occasion for wild demoniacal behaviour, jeering and cursing the dying man, unawed in the presence of death. Then his body was ignominiously cast into the nearest ravine to be food for the hyenas at night.

Not only was the poison ordeal resorted to in cases of supposed witchcraft, but the Tonga and Tumbuka, with whom and not with the Ngoni the practice originated, were incessantly using it. In nearly every hut a bundle of poison-bark would be found hid away in the roof against the need to use it. Family and other quarrels were finally adjusted by resort to the ordeal. The women were the mainstay of the horrible practice, and most frequently made use of it. Numberless cases were treated at the dispensary, when more sober reflection made them seek an emetic. Sometimes cases were brought by others.

A Village Audience.

A husband might come home and find a crowd about his door and learn that his wife had taken *muave*. He would bring her to me at once. Sometimes the patient has died while being brought, or even at the dispensary door while I was making an effort to save her. Frivolous as were the reasons for resorting to such extreme measures when quarrels arose, there were often dire results therefrom, and sometimes one met with a case which appeared ridiculous even to the native mind. A strong young man came to me one day saying he had drunk *muave*, and desired an emetic. On enquiry I learned that he and his wife had quarrelled during the night in the secrecy of their own hut. Failing to agree after the usual amount of talking characteristic of native brawls, they agreed that at sunrise they would drink *muave*. When the sun rose they proceeded to the ordeal and the cups were duly mixed. The wife, with a cunning not suspected by the pliable husband, who, with a faith in his innocence, was determined to go through with the business, said, " You made the charge, so you shall drink first." He did so, but the wife, hurling an imprecation at him, refused to drink her share, and fled to a village several miles away. The poor man, amid a crowd of natives derisively cheering him, came

and sought relief, which a liberal use of sulphate
of zinc and water gave him.

The poison ordeal is an outcome of their belief
in the supernatural. It is an appeal to a power
outside themselves to judge the case, reveal the
right, and punish the wrong-doer. It is part of
their religious system and appears to them to be
right. The witch-doctor is to them the visible
and accessible agent of the ancestral spirits whom
they believe in and worship, and from whom they
think he derives his powers. If there is a tendency to error in what they believe, the witch-
doctor by his shrewdness and making bad use of
it, pretending to know more than what will ever
be revealed to man, favoured the growth of lies,
and juggled with the truth of things. The char-
acteristics of the witch-doctor are a pretended
superior knowledge to discern the affairs of individuals and communities, and ability to hold
intercourse with the ancestral spirits. It is not
a hereditary craft such as that of other kinds of
doctors, *e.g.* medicine men who have a knowledge
of herbs, and blacksmiths who have the secrets
of working in iron. The knowledge of medicine
and handicraft are considered to be heirlooms.
The witch-doctor is supposed to be chosen by the
ancestral spirits, by whom they may communi-
cate with the world. A man who is chosen

presents certain features or symptoms. He becomes "possessed" and excludes himself from society. He may have a peculiar sickness, characterised by lowness of spirits. It may be he is the subject of fits or has peculiar dreams. When he recovers from this and again enters society he is looked upon with awe by the ordinary people. He places himself in the hands of some old witch-doctor who tests his symptoms of "possession," and if found good he is instructed by him in various practices. He is not allowed to graduate, however, until he has discovered some medicine which is potent in some way, and given public proof of his ability to discover things secreted by those assembled to test his powers. There is doubtless a measure of both self-deception and imposture in the matter. The practice of the witch-doctor is closely connected with the worship of the ancestral spirits. Each house has a family spirit to whom they sacrifice, but no one ever sacrifices to the spirit without first waiting upon the *itshanusi*. He pretends to have found out the reason for worship, and directs the applicant how to proceed.

Without asserting that it is complete, the following is a correct statement of the religious beliefs of the natives. Although they do not worship God, it is nevertheless true that they have a distinct idea of a supreme Being. The Ngoni call him *Umkurumqango*, and the Tonga and Tumbuka call him *Chiuta*. It may be that the natives, from an excess of reverence as much as from negligence, have ceased to offer him direct worship. They affirm that God lives: that it is He who created all things, and who giveth all good things. The government of the world is deputed to the spirits and among these the malevolent spirits alone require to be appeased, while the guardian spirits require to be entreated for protection by means of sacrifices. I once had a long conversation on this subject with a witch-doctor who was a neighbour for some years, and the sum of what he said was, that they believe in God who made them and all things, but they do not know how to worship Him. He is thought of as a great chief and is living, but as He has the ancestral spirits with Him they are His *amaduna* (headmen). The reason why they pray to the *amadhlozi* (spirits) is that these, having lived on earth, understand their position and wants, and can manage their case with God. When they are well and have plenty, no worship is required, and in adversity and sickness they pray to them. The sacrifices are offered to appease the spirits when trouble comes, or, as when building a new village, to gain their protection.

With such ideas native to the mind of these tribes, how is it that the materialistic writers and unbelieving critics of Missions affirm that the high moral and spiritual truths of Christianity cannot be grasped by them? In beginning mission work among them, one is not met by anything in their mental or spiritual life which is an insurmountable barrier in communicating to them spiritual truths. However erroneously at first they may conceive the truths and facts put before them, they have no difficulty in finding a place for them in their thoughts. To talk of spiritual things is not to them an absurdity, much less is it impossible for them to conceive that such things may be. The native lives continually in an atmosphere of spiritual things. Almost all his customs are connected with a belief in a world of spirits. He is, consciously or unconsciously, always under the power and influence of a spiritual world. In preaching, we have not first to prove the existence of God. He never dreams of questioning that. We have in our instruction merely to unfold His character as Creator, Preserver, Governor, and Father of us all. As He is revealed to them they do not question His sovereignty, but bow to it. While we meet with many obstacles in their life and thought, yet as they are we have in them much that is a help—a basis on which we may operate. However dim their spiritual light may be, we have but to unfold truth to them and it is self-evident to their minds. No preparation by civilization is required, as their spiritual instincts find in the truth of God what they are crying out for. The cry is inarticulate and unuttered, save in their unrest and blind gropings after spiritual things.

Regarding the origin of life and death, all natives have the story much the same as found throughout the Bantu tribes, how that in the beginning God sent the chameleon to tell men that they would die but again rise. Afterwards He sent the grey lizard to say that they would die, and dying, would not return. The lizard, being a swift runner, came first, and afterwards the chameleon ; but men said, " We accepted the word of the first, and cannot receive yours." The natives hate the chameleon, and put snuff in its mouth to kill it, because they say it delayed and led to their acceptance of death.

They believe in the presence of disembodied spirits, good and bad, having the power to affect men in this world. Their sacrifices to them, their fear of them, and their assigning sickness and death to their agency, testify to this.

There are different terms applied to spirits, each of which is explanatory. The native thinks of the shade or shadow of his departed friend, and denotes the life-principle, and the term is even applied to influence, prestige, importance. They use it in reference to his life, as when they say, "His shadow is still present"; meaning that though on the point of death, his spirit is still in him. When I began to take photographs, the same word was applied to a man's photograph, and they evinced the greatest fear lest by yielding up their spirit to me they should die. I have shown photographs of deceased persons known to them, and they invariably turned away, some even running away in fear. When a native dreams, he believes he has held converse with the shade of his friend. Another term applied to spirits has reference to their supposed habit of wandering about. The hut of a deceased adult is never pulled down. It is never again used by the living, but is left to fall to pieces when the village removes to another locality. They do not think the spirit always lives in the hut, but they think it may return to its former haunts, and so the hut is left standing. Spirits are thought to enter certain snakes, which consequently are never killed. When seen in the vicinity of houses, they are left unmolested; and if they enter huts, sometimes food and beer are laid down for them. Some time after a chief died, some of his children saw a snake near his grave close by the hut in which he died. The cry of joy was taken up by all the family, " Our father has come back." There was great rejoicing, and the family went and spent a night at the grave, clearing away the grass and rubbish that had accumulated. They were satisfied that it was the spirit of their father in the snake.

If a journey of importance is being taken, such as an army going out to war, or a man going on important business, a snake crossing the path in front is considered to be an omen — the spirit giving warning against going on. The army or party interested would not dream of going farther, without consulting the divining-doctor so as to learn the meaning of the omen.

Their belief in spirits appears on many occasions. I have been engaging workers when only a few out of a crowd could be chosen. It was not an uncommon thing to hear from the disappointed as they walked away, " I have an evil spirit to-day," meaning that luck went against them, and they were not engaged. A man who has perhaps narrowly escaped from danger exclaims, " What did they take me for ? " meaning that some inferior spirit had been caring for him, and only barely saved him. Such a definite and operative belief in the presence and power of spirits gives rise to their practice of offering sacrifices, which are almost always propitiatory, save when a new village is made. Hence their religious exercises are called forth by sickness, death, or disaster. A man speaks of a sacrifice as offered to make the spirit pliable and obedient to his request, and in sacrifice they offer cattle, or beer and flour.

Although the Tumbuka are a much more degraded people in morals, they are more religious than the Ngoni, and are freer in their sacrifices. An elephant-hunter, for example, when the beast falls, always cuts out certain parts, and at the foot of a certain tree offers them in sacrifice to his guardian spirit. Their beliefs and worship are essentially those of the Ngoni, except that they have a wider variety of objects. Certain hills are worshipped, also waterfalls, ancient trees, and almost any object which appears unusual, may to them embody the spirit they worship, while certain insects, such as the *mantis religiosa* are supposed to give residence to an ancestral spirit, are not interfered with under any circumstances, or even handled. Each house has its own guardian spirit, and the tribe worships the spirit of a dead chief.

The natives believe in Hades — the region below, where disembodied spirits dwell. They do not speak of it as a sensible locality. Now and again women are found wandering about the country smeared with white clay and fantastically dressed, calling themselves "chiefs of Hades." They are greatly feared as being able to turn themselves into lions, and other ravenous beasts to devour any who may not treat them well. Hence their advent in a village leads the people to give them whatever they ask, that they may go away and leave them undisturbed. There is a medicine in use as a protection from lions, which cunning men sell at a good price. One of the largest and most attentive meetings I had in the open air was when, on a Sunday morning, I came upon a crowd of natives of both sexes and all ages, submitting to be anointed by a deceptive old man with an oily mixture, which was reputed to give protection from the lions at that time infesting the district. At my request he ceased his practice and I preached from the words : " The devil goeth about as a roaring lion seeking whom he may devour." Before the close of the sermon the old man took his departure with his oily mixture, leaving me in possession of the crowd.

Much more might be said of the life of the people, but what has been stated will enable the reader to understand the nature of the soil into which the seeds of Christian truth have been cast, and how great have been the results. Frederic Harrison's New Year Address to the Positivist Society ten years ago contained these great swelling words of man's wisdom :—" Missionaries and philanthropists, however noble might be the character and purpose of some few among them, were all really engaged . . . in plundering and enslaving Africa, in crushing, demoralising and degrading African races." I have but faintly touched upon the moral and spiritual, as well as the temporal state of the natives as we found them ; let the reader, when he has gone through the succeeding chapters, say for himself whether the plan of God's redemption of Africa or that of the Positivist Society succeeds best, and take no rest until all Africa receive the light of God's Word.

CHAPTER IV

THE STATE OF THE COUNTRY AND THE BEGIN-
NING OF THE MISSION WORK

WHAT has been said in the introduction shows the position of the work in Ngoni-land in relation to the more extended operations of the Livingstonia Mission as a whole. In the history of the Ngoni, as given in the previous chapters, we are brought down to recent times, and have now to hurriedly glance at the state of the country produced by their presence and power in Nyasaland at the advent of the Mission.

Soon after the work was begun at Cape Maclear, near the south end of Lake Nyasa, it was evident that if the Mission was to be established according to the idea of the pro-moters, a wider and healthier area must be found. To secure that different expeditions were undertaken, and it was in connection with these that the full extent and power of the Ngoni became known. Reference is made here to the reports of these expeditions by Drs Stewart and Laws, and the late Mr James Stewart, to the Royal Geographical Society, and to the Committee of the Free Church of Scotland. One of the earliest references to the power and dominion of the Ngoni, over a wide area, was made by Dr Stewart in the Free Church General Assembly in 1878, when he said, regarding the position of the newly-formed Mission to Nyasaland, "He had recommended a change of site, and preparations had been made for carefully examining the portion of the country on the western side of the lake. There was a certain responsibility in connection with this recommendation to change the site. He was willing to face the responsibility. They had either to make a change or let go the original idea and projection of Livingstonia, and reduce the whole to dwarfish proportions, very different from what was at first intended. What was urgently wanted was a high and cool position possessing all the other qualifications and capabilities of a good site. These he thought might be got on the high lands to the west side of the lake. The warlike Ngoni were in possession of that district. If we could establish friendly relations with them the work would not be difficult."

Thus it was, twenty years ago, that in the

providence of God, through the contracted area workable from Cape Maclear as a centre, and its unhealthiness, the Mission was led to interest itself in the proud warriors of Ngoniland far away to the north among the high hills on the west side of Lake Nyasa. The reputation for war and cruelty which they had wherever they were known, made the task of finding a new site anything but easy, notwithstanding the hopefulness of Dr Stewart's report. The following account of what was found during the expeditions undertaken, and of the origin and progress of the work in Ngoniland, should be read with interest in view of the now enormous field of the Livingstonia Mission, and the wonderful achievements of the Gospel among many different tribes.

The Ngoni at that time dominated a tract of country extending between 9·30° and 12° S. lat. and from the western shore of Lake Nyasa to 31° E. long., comprising an area of 30,000 square miles. In this vast region the principal tribes living were,—on the lake shore, the Tonga, Tumbuka, Henga and Nkonde, while inland were the Chewa, other divisions of the Tumbuka, the Senga, Zingwa, Wiwa, Bisa, Nyamwanga, Wanda and Nyika, and other communities which were scattered remnants of tribes already broken up by their arms. When it is remembered that every year during the dry season, which extends from April to November, the Ngoni armies were engaged in raiding expeditions, sometimes to the southward against the quiet and industrious Chewa, or down to the lake shore against the Tonga, or northward to the cattle-keeping Nkonde, or westward into the land of fat sheep, ivory and copper wealth, going as far as Bangweolo, near the site of Livingstone's death-scene, it may be imagined that the condition of these people was anything but happy or secure. I have seen an army, ten thousand strong, issue forth in June and not return till September, laden with spoil in slaves, cattle and ivory, and nearly every man painted with white clay, denoting that he had killed someone. Around Bandawe, one of the principal stations of the Mission, more blood has been shed than can be related. The Tonga, once enslaved by the Ngoni, but who revolted and fled, were the frequent objects of attack. Ngoni wars, notwithstanding the reputed bravery of the warriors, were not always very straightforward fights, but were always very bloody from the tactics they pursued. The army would lie concealed in the forest at some distance from the lake villages, and when the sun was dipping behind the hills it would rush out and enter a village at a time when all were congregated and engaged in the open air. It was but a rush through the village, but ten, twenty, or thirty men, women and children were left lying dead, and perhaps as many women carried off captive.

I was at Bandawe when such an attack was made on a village a few miles from the station. We were seated in the verandah of the Mission house in the calm, cool evening. The boys boarded on the station as Mission pupils were engaged in mirthful games near by. In the villages around, hidden among the banana groves or rich undergrowth, we could hear the thud of the pestle in the wooden mortar as the women, with their babies tied on their backs, were employed preparing the flour for the evening meal. The children were heard in gleeful song and dance, while the hum of voices rose as the men engaged in the gossip of the hour, seated under the village tree smoking their pipes, and the sun sank amid a splendour of colour over the western hills. All betokened peace and happiness. Suddenly a shrill cry was heard in the distance and it was at once taken up by those in the villages, the song and gentle hum of voices giving place to cries of fear and distress. Before many minutes had elapsed hundreds of frantic women carrying their infants, while older

POKA HUTS ON HILLSIDES.

children ran frantically by their side, rushed into the station grounds or off to the caverns on the rocky hill near the shore. The men fled for their arms and soon the tumult of battle was heard. An Ngoni army had rushed a village; the peace and quiet of the evening hour now gave place to the wailing of women and the cries of children, as they re-entered their villages to find perhaps several of their friends killed or carried away captive. On one occasion such an attack was made and several women were carried off. Some men who had guns went in pursuit and traced the route of the Ngoni by the bodies of the dead whom they had slain on the way, finding they were not after all worth carrying off. Coming up to them at a river where they were encamped, still having in their possession some women, they surprised them by firing their guns. The Ngoni fled, but one woman, about to become a mother whom they could not urge to run, was speared to death before the eyes of her friends who had come to rescue her. I have seen an infant with a great ugly gash in its little body which was made by the spear that passed through the mother as she rushed off in the effort to escape.

The following is also an authentic story of an Ngoni war and butchery told by a European

F

who witnessed the sight, and such harrowing
tales could be multiplied tenfold.

" On Friday, Nov. 18, a band of Ngoni
stealthily surrounded the village of Kayune
which lies on the lake shore. They had no
dispute with chief or people ; their one motive
was to spear men and capture women. There
was no moonlight and darkness favoured their
approach. Entering the village, which had no
stockade and lay half hidden in banana groves,
each warrior took up his position at the door of
a hut and ordered the inmates to come out.
Every man and boy was speared as he rushed
out and the women were caught and bound with
bark rope. In the morning not a Nkonde man
or boy was in the village, while three hundred
women and girls were tied and crowded together
like so many frightened sheep. The Ngoni
feasted all day on the food and beer of the
villagers."

The sequel is, if possible, more horrible. A
party of traders at Karonga, three hours' journey
distant, went out to try and rescue the women
when word of the capture was received. The
party came up on the Ngoni and fired upon
them. They were off their guard and supposed
that a large force had come against them, and
they began to spear their captives. The writer

goes on to say, " Then ensued a horrible scene,—
women screaming, women wrestling for life with
armed savages, women and girls writhing in
blood on the ground." Eventually two hundred
women were rescued. The number killed in-
cluded twenty-nine men, one hundred women,
thirty-two girls, sixteen boys.

No one can estimate the loss of life in peaceful
tribes, or measure the anguish and distress, not
to mention the incessant state of fear, in which
these tribes lived, due to the position and war
power of the Ngoni. When Dr Laws and Mr
Stewart passed through the country in 1878, in
pursuance of their search for a new site for the
healthy station already referred to, they everywhere met with traces of the Ngoni power and
cruel wars. Along the lake shore they found
the people compelled to live in swamps amid the
stench and death-dealing exhalations, struggling
for an independent foothold on mother-earth,
in some of which I have had, in carrying on
medical and evangelistic work in that district, to
be carried from door to door on a native's back
as the paths were all under water, or semi-liquid,
black, stinking mud. In other places they were
to be found crowded together on some neck of
land or secure place surrounded by a triple
stockade of strong trees. Dr Laws mentions

one such, near what is now Bandawe station, where the Chief Marenga (who now lives in happier times in an extensive open village) had a triple stockade round his village, the distance between the stockades being from 30 to 60 yards and the interval filled by growing jungle. At another place it was said, " The people here might be said to be almost driven into the lake by their relentless foes, the Ngoni. The stockades ran 30 yards into the lake itself, and the greater number of the huts were actually built on the sandy beach." Even as far north as Karonga's at the north end of Nyasa, they found the dread of the Ngoni pervading the community, and the old chief made a present of a young bull and a tusk of ivory to Dr Stewart to induce him to give him medicine to fight the Ngoni. Along the lake shore, north of Bandawe, the hills dip down with precipitous sides almost into the lake, and the shingly beach was occupied by villages where there was some degree of safety. They managed to barely exist by planting patches of cassava where any soil could be found on the crags above, the people not daring to go far from their homes. In the lake, towards the north end, there are rocky islands. They are huge accumulations of boulders—as if they had gradually grown out of the water by added masses—on which there was little foothold or place to make even a hut such as the natives usually build. Yet on such islands scores of poor Tonga, Tumbuka and Henga, had their only sure place of abode. Driven off the face of the earth, as might literally be affirmed, they had to rear their families, cradling them in the cracks of the rocks or crannies between the boulders, to prevent their rolling off into the water. The only shelter afforded was by making wattled shades over which a few handfuls of grass were laid to protect them from rain and sun. When they considered it safe they would paddle their canoes to the shore, and snatch a few hours' work in their patches of potatoes or cassava and betake themselves again to their rocky home.

Again, high up on the most inaccessible parts of mountain ranges, the remnants of broken tribes, and even whole tribes, had their dwelling. They had their grain-stores hid away in the darkness of the remnants of the great primeval forest still met with in the ravines on the mountain sides. Their dwellings were in some cases no more than a hole scooped out on the bare steep side of the mountain, and a few sticks pushed into the earth above projecting over the levelled spot, with a little grass over them. The best of them was of the rudest description, while all around the ground sloped so sharply that one could not walk without holding on to objects. Their crops were peas which they cultivated on the declivities, by sowing rows among the bracken which they left as supports, and to prevent the soil from being washed away in the rainy season. The ingenuity of such a people in providing themselves with the bare necessaries of life, could scarcely be admired properly, from the sad feeling at the thought of how they had been hunted and reduced to such a condition. On the approach of enemies they fled into the dark forest and had nature for a guard. Wherever on the mountain slope, at Mount Waller for instance, space whereon to erect a hut could be found it was utilised. Lying on board the steamer in Florence Bay with the vast pile of that mountain before us, the terraced slopes were seen to be crowded with huts, a situation from which no Ngoni army could dislodge them. One of the most remarkable sites for human habitations was found at Manchewe in the neighbourhood of Mount Waller in 1895, when Dr Laws and I were examining the district preparatory to founding the Livingstonia Training Institution.

On ledges of rock on the face of a cliff 250 feet high, a section of the Nyika tribe had their homes. Over the cliff, 200 yards apart, two rivers poured their waters in a series of waterfalls into the wooded gorge below. The face of the cliff was covered with a profusion of tree ferns, magnificent aloes in bloom, many beautiful ferns and other tropical plants, from among which tall, graceful trees sprang. A full description of such a combination of natural grandeur of rock and tree and waterfall is impossible. Here we want merely to picture human beings living between earth and sky in small circular huts, in some cases built on ledges of rock not ten feet broad, and in other cases, the houses being actually tied to tree roots which have, in growing, split the rocks, and in some cases dislodged great masses. The overhanging cliffs and mighty trees above, with the depths below, formed the natural protection to that poor hunted people. Access to the villages was had by scrambling down the fissures in the rock, or by hanging on to tree roots or other projection which would afford help and safety. The clusters of huts were partly hid by the dense undergrowth, and only those guided by the natives could have found the safe ledges along which to pass. Viewed from above one was forcibly reminded of the home of the sea-fowls on the cliffs around our coasts.

In no case was the rocky ledge on which the houses were built more than twenty feet broad, and it made one shudder to look down on the little children playing around the small huts, with the roaring cataract at one side and a sheer precipice above and below. In time of war, or danger from falling rocks dislodged by the rains, the caverns found near were the hiding-place of these inhabitants of the rock. Their homes were made seemingly in defiance of nature's great law of gravitation, — forced over the edge of the world, so to speak, by the inhumanity of the Ngoni. If the Gospel can do anything at all to better men's lives, there, surely, we found a fit field for it.

Great must have been their surprise when they saw many of their Ngoni enemies standing on the heights above, calling to them that they came on a peaceful errand, and inviting the men up to speak to us. We arrived on a Saturday evening, and having made friendly overtures, we invited them up to our camp next day to join in the worship of God. For the first time in their lives and in that district, the voice of praise and prayer was heard, and these wretched people heard in the Ngoni speech the word of peace and not of war. Surely that day in that place the prophecy was fulfilled : " Let the inhabitants of the rock sing, let them shout from the top of the mountains." There came to them that day the dawn of a better life, as we shall see in due course.

While such a state of terror and distress was known to exist over the country lying between Ngoniland and Lake Nyasa, there remained the vast country unexplored, lying to the west and north of Ngoniland, upon which horde upon horde of savage Ngoni waged a relentless war. The same state of terror and distress obtained there, but was known only by the spoils brought back by the armies. Mr John Moir in 1879 made a long journey into that region, and everywhere saw evidence of the Ngoni raiding. Later on several Europeans passed through the district, and all met with the same story of Ngoni wrongdoing and domination. Last year a careful survey of the district was made by my colleague, Dr Prentice, to find out suitable localities for new stations, and I heard him relate in a public meeting at home, an incident which may fittingly find a place here as bearing upon the past condition of the people all over that region.

In the course of his journey he came upon a considerable community huddled together in poor houses, in the centre of a great swamp, through which he could not find a way. The

village was also strongly stockaded, and it was
evident that they had recently rebuilt it.   Finding it impossible to enter, he fortunately saw
a native who had been at one of our northern
Mission stations, and could understand what was
said, and the object of the expedition.   By him
communication was had with the people in the
stockaded village, and Dr Prentice and his Ngoni
carriers were invited to enter.   In conversation
with the affrighted natives, the chief said that
long ago they were hunted by the Ngoni, but
that in recent years they had heard of men
coming to them with a book which they had
accepted, and had consequently given up war.
Recently, however, they had heard that Mombera
the chief had died, and on the placing of a new
chief they feared that the Ngoni might again
break out, so they had taken the precaution of
removing their chattels to safe quarters to await
the attack which they apprehended.   One of the
Ngoni carriers thereupon took from his pocket a
copy of the Gospel in Ngoni, and declared that
now the Ngoni had accepted the book, so that
they need no more fear an attack, and he added,
"Long ago we came with war to destroy, but
to-day we are one with the white teacher, and
come to bring you good news of peace and salvation."   To have witnessed such a scene more

than repaid Dr Prentice's weariness and sickness on his long and trying march.

The only tribe that withstood the Ngoni was the Wemba to the south of Tanganyika, and many and fierce were their contentions. The picture of Ngoni power and incessant raiding is complete when I add that, in Ngoniland, there are representatives of at least sixteen different tribes found among their slaves, their original homes lying in the region from what is now the Colony of Natal on the south, to Tanganyika on the north, Nyasa on the east, and Bangweolo on the west.

Such, then, was the character and such the reputation of the Ngoni, when the Mission pioneers first met them. In 1878, Dr Laws and the late Mr James Stewart found a probable site for the new station on Mount Kaning'ina on the outskirts of Ngoniland, and between it and the lake. Here for a time the late William Koyi (the Kafir member of the staff, to whose life and work a special chapter is devoted) and a European were located to observe the nature of the district, and, if possible, to become acquainted with the Ngoni. They managed to form an acquaintance with a Swazi family—the Chipatula family—living not far off, and through them obtained an introduction to Mombera, the chief of the Ngoni, and to Mtwaro, his brother and successor. The Chipatula family had been at one time strong in power, and to them belonged most of the Tonga who revolted and fled to the neighbourhood of Bandawe station, subsequently chosen instead of Kaning'ina. Dr Laws realised the nature and difficulties of the task set him, although the suspicions of the Ngoni were aroused and persisted for many years, by the Mission having located itself among the Tonga slaves on the lake shore, and having for a time occupied the outpost at Kaning'ina as if to set a watch upon them. Yet the wisdom of the step, and the caution necessary in every movement, have been fully justified in subsequent years.

William Koyi had been able to find out the Ngoni centre of power, and to be received by Mombera in a friendly manner. Between Dr Laws's first and second visits to Kaning'ina, however, a rising took place among the Tumbuka and Tonga slaves in Ngoniland, which at the time threatened to destroy all hope of access to the Ngoni. They believed that the freedom of man which the Mission expedition, with its retinue of native servants and carriers belonging to different tribes, embodied, had emboldened their slaves to revolt. Many of the Tonga fled to the lake shore, but the Tumbuka, less successful in their effort to escape, were forced up Hora mountain (where now one of our Ngoniland stations is situated), and starved into surrender.　Being allowed to come down to drink at the fountains around the base of that bare, rocky height, the Ngoni fell upon them, and many hundreds were massacred.　I have seen the skeletons lying crowded together around the foot of the hill, and also upon it, some being found in caverns and at the foot of precipices where they had been slain.

We need not describe in detail the transactions between the Mission and the isolated workers holding the outpost at Kaning'ina—a situation often fraught with great personal danger through the opposition of the Ngoni and the treachery of the Chipatulas, who all the while pretended great friendship.　The fact was, the Chipatulas were diminished in power and influence among the Ngoni, and hoped, by means of friendship with the white men, to regain their power.　An example of their duplicity we find in the statement of the dead chief, Chipatula's brother, when Dr Laws first met him.　He was asked about the chief, and his reply was that the chief was dead, and that until he (Chisevi, the speaker) should go out of mourning and be crowned, Mombera, a *headman*, was ruling, whereas at that time the chief over all the country was Mombera, and the Chipatula family were ordinary members of the tribe.

In 1879 Dr Laws first met Mombera. Ever since then Mombera shewed a strong affection for Dr Laws and unbounded confidence in him, and through him as " the father of the white men," in those who followed him and lived under him. The only parallel to this mutual regard which I know is in the case of Moffat and Umziligazi, the *confrère* of Mombera's father in the far south in the beginning of the century, as recorded in " Robert and Mary Moffat." Here we have two bloodthirsty, despotic chiefs, far apart but of the same blood, visited by two missionaries of the Cross, and without in the least degree to all appearance accepting any of their teaching, forming a strong attachment to them, and till death maintaining it, and speaking often of it. Living, as I did, with Mombera for six years before he died, I never knew of his having stopped a single war party from attacking the helpless Tonga around Dr Laws's station at Bandawe because of his belief in God ; but over and over again, because of his attachment to Dr Laws, he refused to sanction war ; and to-day thousands of Tonga men and women owe their life to Mombera's affection for Dr Laws. Happy, indeed, must he be who was thus used of God in saving the lives of so many people, that they might, as they now do, hear and receive God's Word.

But to proceed with the narrative. On the 24th January 1879 Dr Laws arrived at Mombera's town and pitched his tent. Of his interview with Mombera, he reported: "We explained to him that the object of our coming into the country was to be friends of all the people, to teach them about God and what He has done for us; that we also wished to teach children, so that they might be able to read God's Word for themselves. We showed them a Kafir Bible, from which William Koyi read a few verses. We introduced Mr Moir as one who was ready to trade with them if they desired to do so, and who loved God's Word as we did. We gave Mombera a present of various articles, with which he expressed himself very much satisfied. The head councillor of the village answered for him that they were glad of our visit, and that they were willing to be friends, and thanked us for our present. They expressed their disappointment that we should remain among the Tonga on the Lake shore, or even at Kaning'ina. 'Why,' said they, 'do you not come up and live with us? Can you milk fish that you remain at the Lake? Come up and live with us and we will give you cattle.   We are the rulers ; the Tonga are under us, although they have broken off from us at present, and run away with our children ; we wish you to make them send back our children.   They say they do not like to live with us because we are cruel.   We are cruel, but not to our children, only to those against whom we go to fight.   Our children we must have back, and we would have gone and fought with the Tonga, and driven them into the Lake, had you not visited us and said war was bad.   We have been defeated ; but when we set about fighting, we do not give up our object, though the last Ngoni should be killed. You say there should be peace ; send back our children and there will be lasting peace.'

" We explained to them that our commission was to bring the Gospel to every creature, to the despised Tonga as well as to the Ngoni themselves ; that we required to have a port on the Lake, so that we might get a supply of provisions, calico, etc., from home, and this we could not have if we were living with them while the Tonga between us and the Lake shore were our enemies ; that we showed our desire to benefit them by not confining ourselves to the Lake shore alone, but establishing a station at Kaning'-ina, near Chipatula, where they could easily learn about us at any time, and have the false

reports they heard about us from the Tonga rectified by a visit. Regarding the sending back of their children we explained that we had not come to interfere with their quarrels, but that we were willing to do what we could as peace-makers, and advised that they should have patience and live in peace, as the best way of having their children brought back to them; and this we considered it to be their duty to do, seeing that the Ngoni had been the original invaders of the country and the disturbers of the peace. They asked that, if a white man could not be left there, one of their own tribe should be sent to them. We told them that in course of time we would endeavour to send a teacher to them also. The chief sent us a small elephant's tusk as a present, and sent a calf to our tent as food. He also sent a small tusk which he said he had intended as a present on our first visit."

On a subsequent visit paid to Mombera by Mr Stewart, he refused to see him, being displeased that the Mission should have visited other tribes first. It was evident that the Ngoni desired an exclusive alliance with the Mission, and, as will afterwards be seen, this idea led to frequent trouble, at times great and prolonged. In the end of that year Kaning'ina observing station was given up and Bandawe founded. Dr Laws

very properly gauged the situation when he
wrote in October : " I do not think it would be
advisable to continue the station in that district
meantime.   More good could be done by pushing
it forward into the country of the Ngoni."

Two years elapsed before this could be accomplished, during which time all round Bandawe
station Ngoni raiding went on. These raiding
parties did not always represent Mombera's army,
but were bands of wild youths who were eager
to obtain wealth, or wives and slaves, and
frequently were led by members of the villages
which they attacked, and who to revenge some
wrong done by their chief went up to the Ngoni
and formed a league with them. Reading
through the journals of Bandawe station in
those days we continually meet with references
to Ngoni raids on the surrounding villages, and
the continued unrest of the Tonga, which rendered the work of the Mission futile, and created
difficulties and dangers in living among them,
as they clamoured for Dr Laws and the Mission
party to join them in fighting the Ngoni. It
was no easy task to live among them and declare
inability to help them in righting their wrongs,
and to have the nature of the work misjudged,
by expecting and demanding temporal good by
force of arms.

For example, on November 17th, 1881, the following entry occurs : " Chikoko came to-day asking about the Ngoni and what was to be done. He described them as a wild beast, and said, ' You cannot hold discussion with a wild beast, you must go to him with a gun. The Ngoni are like a snake, we, like a frog. When a frog sees a snake he goes off hop, hop, hop, to save himself. That is how we do, and the people are leaving their villages and coming to the beach all round.' He asked for guns and powder. This Dr Laws at once refused to give him, and told him we had no intention of fighting with the Ngoni. We brought the Gospel to the Tonga, and meant to take it to the Ngoni, and if we fought with them it was not likely they would be willing to receive it. ' But,' said Chikoko, ' they will kill you, and destroy your goods.' Let them destroy them if need be—God will protect us."

" Tuesday, Dec. 6. In the morning a report reached us that a party of Ngoni had made a descent on the Matete valley and had killed five men (three by another report), and had made *misasa* on the east side of the stream. In the forenoon Chikoko, Chimbano, Katonga, Marenga, Mpimbi, Marengasanga, and other chiefs assembled, wishing to have a consultation with Dr

Laws. Dr L., S., and M^cC. heard them. They
said the Ngoni had come down bringing a man
who had come from Matete, having made a
narrow escape from the hands of the Ngoni who
had surprised him and his companions while at
work in their gardens. The chiefs asked Dr L.
what they should do. Dr L. reminded them
that it was not his work to settle their disputes,
and that they must consider with themselves
what they should do, and do as if there were no
English here. But before, Dr L. was willing to
go and speak with the Ngoni, would he not go
to Matete to-day and see them ? Dr L. replied
that that last time when he was ready to go no
one knew exactly where the Ngoni were, and
to-day he was busy with other work. Dr L.
thought it very probable that should he go he
might be accompanied by a great many more
than he would desire, and that they would be
anxious to begin a fight with the Ngoni in which
he would be implicated. Chimbano said that
they were now hearing God's Word and obeying
it : that we had told them war was bad, and that
they should not sell people. They did not want
to fight but live in peace, and here were the
Ngoni coming and killing them, if Dr L. waited
till the steamer came with those whom he expected to go on to Mombera's, they would all be

killed, and the white man did not want to live in the wilderness without them. As for the Ngoni, they were too wicked to receive God's Word. All the villages of the Tonga for many miles north and south had been destroyed by the Ngoni, only Chintechi remained—Mankambira and Kang'oma had sold some of their people for guns, and now they were able to repulse the Ngoni, so they had better do the same.

" Dr Laws reminded them that a few months ago Chimbano had gone and made war on Mankambira when told he was doing wrong, and that such was a strange way of obeying God's Word, and if he chose to sell his people for guns, the *mrandu* would be between God and him. Dr L. further said, ' You want us to go and fight the Ngoni.' Yes, that was the very thing. Well we are not going to do it, we have told you so before, and we tell you so again. When we came here we told you we were not to take part in any of your quarrels and fight for one side or the other. We have orders to this effect from home, and Christ has commanded His Word to be taken to all nations. We went before to Mombera and made friends with him. We do not wish to fight against him, nor against you, but to teach all. The Ngoni have received the Word of God in the south and may do so on the hills here, but it is not likely they will receive us with it if we fight with them here.' There was a good deal more talk to the same effect, but not being able to change our intention they showed their spite by calling off all their people at work on the station, and issuing orders that the first one found working with the English was to have his house burned down. In the afternoon only two or three of the Chewa and Tumbuka tribes were found working. In the evening many people assembled, armed, and marched by moonlight towards Matete. Last night a watch was set and two men were detected in an attempt to open the byre and fled. The watch set again to-night as Tonga movements might be as hurtful as Ngoni ones."

The next approach to Mombera and the Ngoni occurred as described in the following entries in the Bandawe Journal :—

"Tuesday, Jan. 10th, 1882. To-day William Koyi with Albert and Jodi and carriers of goods started for the hills to visit Mombera, going first to the village of the Chipatula family."

"Jan. 25th. Albert returned from the hills to-day, bringing a letter from William Koyi. They report great scarcity of food among the Ngoni. . . . William Koyi has not yet seen Mombera, but he has had communication with

Ng'onomo. Many of the people were favourable to us but many were inclined to show hostility."

A temporary peace between the Ngoni and Tonga was at this time established. William Koyi took possession of Ngoniland for Christ, and inaugurated a long period of waiting ere the chief and his headmen permitted the work to be fully carried on. Dr Laws also visited Mombera that year, and again in 1883, but despite earnest entreaties no permission could be got to open schools, and in Chipatula's village alone was preaching allowed. In the end of 1883 the Ngoni broke the peace they had agreed to, and attacked Fuka's village near the Bandawe station, and burned down the Mission school which had been erected there.

Such were the Ngoni and their neighbours at that time. War, bloodshed, famine and death, with untold misery among those spared, was the condition of countless thousands over the region raided by the Ngoni. But a great forward movement had begun in the Livingstonia Mission, by the building of a wattle-and-daub hut near Mombera's head village with the determination to stay until expelled, full of faith that one day the Gospel would win its way among the people and become the bond of unity between bond and free, raider and raided, in Ngoniland, and in the regions beyond. In the end of 1882, Dr Laws wrote about it :—

" Aug. 30, 1882. William Koyi is doing a noble work among the Ngoni which no European could have accomplished. The people are jealous and conservative in the extreme, and by no means ready to credit disinterested motives in others. William, by living among them, has already to a great extent disarmed their suspicions. He is respected by all, and I think enjoys the confidence of Mombera, the head chief. General liberty has not yet been accorded to us to preach, but public opinion is rapidly moving in that direction, and it only awaits the decision of one or two of the head men of the tribe to make the length and breadth of the land free to the Gospel. Schools are at present prohibited, but even with regard to this a change is coming over the people so that liberty to teach the children may next be expected. Much hard work will have to be done, but that is nothing, if the tribe can be won for our Lord. The necessary basis of the work is the good-will of the people, and I think this foundation is being surely laid."

CHAPTER V

FIRST VISIT TO MOMBERA

CAPTAIN BURTON says, "It is always a pleasure, after travelling through the semi-republican tribes of Africa, to arrive at the head-quarters of a strong and sanguinary despotism." Only those who have lived in Africa can understand how it is so. Journeying inland from Quilimane, I passed the then powerful Matshinjiri tribe at war with the Portuguese on the Shire river, and met a detachment of their army under the famous Raposo, a man of great dignity and valour. Further on I passed through the Makololo remnants of Livingstone's caravan, established as the powerful chiefs on the Shire river. These were fine specimens of humanity and raised one's enthusiasm for work in Africa among such noble people. Getting up to the highlands above the Shire, and meeting with the very mixed people, "a people scattered and peeled," it became at once evident that slavery and war had crushed the spirit of the remnants of

Yao and Mang'anja peoples living there. Every tenth man one met might be set down as a chief, and the usual results of a few people and many chiefs were very evident. The life of the people seemed to consist in a talking *mirandu*. Petty quarrels of petty chiefs were abundant, and those Europeans who had people living on their ground were oppressed by their attempts to settle their quarrels. It was a thankless business, certainly, where the people delight in talking, and can conveniently keep the questions open over many years and even for generations. For such people one of the greatest blessings which have come to them in recent years is that the British Government has become their chief and united all.

Further on one was able to find at Mponda's a powerful chief, and from a native point of view a happy and prosperous people. Around Bandawe, again, almost every village had a distinct chief, and as one of these was sure to be trying to become paramount, petty quarrels and wars were common. Though all were of one tribe in reality, there was no union among them, even against their common foe, the Ngoni. Had the missionaries engaged to settle disputes no other work could have been done; they wisely espoused no one's cause, but remained the friends of all, had access to all, and saved disaster to their work. The bane of the district was the multi-tude of petty chiefs, and they were thus an easy prey to the ravages of the Ngoni war parties.

But it was when I came into Ngoniland that something like Burton's feelings were experienced. The great expanse of country—hills and valleys and plains—dotted over with numberless villages built without regard to safety from attack, but located where the best gardens and pasturage were to be had, made one realise that here was a people powerful and free, whom to settle among, and win for Christ, was a work worthy a man's life. Elsewhere I saw the people huddled to-gether in small, dirty, stockaded villages, the sites of which were frequently found to be surrounded by marshes in order to give protection with the least amount of work on fortifications, and the people of one village ready to make war on the next village a few yards off. But here in Ngoni-land there was one royal residence, one ruler and he in touch by means of the head-men in the different parts of the tribe, with all the people under him. Standing on the hills on the eastern boundary of Ngoniland, and having pointed out to me the various sections of the tribe all under the one chief, Mombera, I remembered the remark of a member of committee when I was leaving home. He said : " If you have faith and patience

to work and win the Ngoni, you are going to the
finest field in Livingstonia." The full truth of
that remark is only now becoming evident.

But now to my introduction to the chief and
his advisers and head-men. No one who has
visited Mombera at his home will forget the discomfort of the ordeal. I had been duly warned
as to his piercing gaze ; his questions as to age,
family, and whether married or single ; his criti-
cisms of one's personal appearance ; and, what
would never be wanting, his barefaced begging
for whatever he might fancy at the time. So,
to have the ordeal past, I set out with Messrs
Koyi and Sutherland to visit Mombera. He was
not in his customary place in the cattle kraal,
but we found him in the small house where he
received visitors and heard cases pleaded when
he was either too drunk or disinclined to go to
the kraal. The hut was enclosed by a neat reed
fence, the space within being smoothly beaten
down and scrupulously clean. Here we found
several parties, who were no doubt waiting to
plead some case before him, and not a few
hangers-on looking for the crumbs which might
fall to their lot when the beef and beer on which
Mombera subsisted were brought in. Having
taken a present for him, I found that several
of his wives were attracted to the place in hope

of sharing the same with their lord. No sooner had they seated themselves and saluted the stranger, than a loud voice, half angrily, half jokingly, asked them what they wanted, and ordered them to be gone. Mombera, with nearly thirty wives, evidently had not a plethora of devotion for them. He said, "You have seen the white man with his bundle, and you come here expecting something. I am here every day, but you leave me alone if there are no goods to divide."

When we had been invited to enter the hut, we did so by going down on our knees and crawling in through the doorway, which was only a couple of feet high and about the same in width. As each entered, the royal salute had to be given by raising the voice, and saying, "Bayete." The joker of our party, who was evidently on very familiar terms with Mombera, shouted, "Be quiet," which was not objected to. On entering the hut, it was some time before the eyes became familiar with the semi-darkness, and then what one saw did not betoken much splendour of royalty. The hut was a round, low-roofed erection, with a well-laid and polished floor of clay. In the centre a round depression in the floor contained the fire composed of logs of wood. To the right of the doorway, on a reed mat, sat Mombera

himself. Beside him was a huge pot of beer, with
a calabash ladle, over which one of his wives pre-
sided, and tempered the beer with hot water. A
smaller pot, made of grass deftly woven so as to
be quite water-tight, was held by Mombera, who
took frequent draughts, and sometimes handed it
round to the people in his presence. If he did
so, or rose from his mat, all shouted, "Bayete."
When he received back the pot, or came in and
sat down, the company shouted, "Bayete." If
one rose to go to another part of the hut, or to
leave the royal presence, he shouted, "Bayete."

To describe the royal dress is not a difficult
matter. The chief part of Mombera's dress was
the numerous beautiful ivory rings which he wore
on his arms, and the rings of plaited brass wire
on his legs. In his ears he wore the usual heavy
knobs of ivory, about an inch and a half in
diameter, and his clothing was completed by a
few yards of coloured calico, carelessly thrown
over his limbs as he sat, consuming his beer or
talking over the cases brought to him for judgment. When not in state at home, his clothing
consisted usually of his leg and arm ornaments.

It was to a new-comer a strange and trying
ordeal to have to sit and be stared at by
Mombera's one eye visible over the beer-pot;
to know that his remarks about one's appearance

were causing amusement to all in the hut, and not to be able to speak, or, indeed, to have permission to speak ; for until one has been greeted by the chief, he must be silent. It was the custom for the chief to refrain from greeting one for some fifteen minutes after he came into his presence. This was considered the best welcome to give, and however trying to one's patience, it had to be borne. On one occasion, when my wife and I had gone to visit a head-man by invitation, we were kept sitting at the kraal gate for over an hour before he came to greet us, and point out a place whereon to pitch the tent. I knew it was the custom to delay thus, and on speaking about it to our host, he said, "Why should I be in a hurry when you come to stay? If a man comes to your house, and you instantly say, 'Good-morning,' that would mean, 'We have only hunger here, so I need not delay you. You may go.'"

The Ngoni salutation is "Tikuwona," "we see you," a slight variation from the Zulu which is, "We saw you." When Mombera had greeted us thus, all in the hut were then free to do so too, and one after another did so in a graceful manner, and to each the proper reply was "Yebo," signifying "Yes." Immediately the tongues were loosened and Mombera plied his enquiries, and passed his judgment on me. Comparisons were made between Mr Sutherland and myself,—who was the elder, were we brothers, why had we straight hair of the same colour, when did we come out of the sea? for the natives thought the white men were spirits who had left their proper dwelling in the water to come and trouble the people. When a convenient opportunity could be got, Mr Koyi informed the chief who I was, and that I had come to ask permission to stay in the country to teach the people the Word of God, and, being a doctor, that I would attend to all who sought help and medicine.

On this an old toothless man, who may be called the chief's mouth, repeated Mr Koyi's statement to the chief. Then the chief replied and his words were taken up by the "mouth" and repeated to Mr Koyi. They were to the effect that he himself was only the chief and the country did not belong to him but to the people. If his head-men agreed to my staying among them he would be very glad and would not offer any objections. He was thereupon thanked for his words and requested to call together his counsellors so that I might meet them and get their permission to stay. This he promised to do at an early date.

On his rising to leave the hut all shouted
" Bayete," and when he was outside a rush was
made by those present for the beer-pot, and a
hearty draught was taken. When Mombera
entered he accused them of having drunk his
beer, but no one of course had touched it—who
indeed would dare to touch the chief's beer, and
who of those present had need to steal, when they
were already bursting with what he had so freely
given? The one predominant feature in native
life is the flattery and insincerity of the people.
In the chief's presence it reaches a climax.

The present for Mombera consisted of some
coloured calico, brass wire, beads, and a few
trinkets such as would please children at home.
He looked at it and demanded a kind of bead
of which we had none. With the most barefaced impertinence and incivility, he replied
saying he would not like to insult the new
white man by refusing what he had brought,
but as there was nothing to be seen, he would
ask me to bring something with me another day.
The trinkets, however, took his fancy and he
adorned his " crown " with some small lockets
and chains, and handed the other things to those
who were in the hut.

Leaving the royal presence, not very favourably impressed by Mombera and his drinking

and begging, I was conducted to the seraglio where the numerous wives of the chief were lying about sunning themselves, or were engaged making beer or cooking meat for their despotic lord. Each greeted the stranger and begged for cloth, beads, and brass wire. Idleness seemed to be the bane of the women and one can imagine that many quarrels and jealousies would arise, demanding the attention of the queen or head wife, whose sphere it was to rule the harem and regulate the number and position of the wives which were constantly being added to, or put out of the way. Mombera had his favourites; these improved their chance and sometimes inveigled him into a union with some near relative of their own. His wives were distributed among his principal villages, either as properly dowried wives, or as the handmaids of such to do their work, and be ready to entertain their husband and his guests whenever he happened to reside at their village. This custom of having several establishments kept up, is the only valid excuse I could ever get for the practice of polygamy. A man would say, " I have gardens and a village at so and so, how can I have only one wife? Who will cook my food and hoe my gardens there ? "

The lot of many of Mombera's wives, and of

many wives of others, was not altogether a happy one. In one instance a principal wife—the chief wife in fact—was slighted by Mombera for some reason and was discarded altogether, and only on his death could anyone be got to espouse her cause, and to put her in her proper position. In another case a wife residing at a distant village at which he had not lived for several years, was, rightly or wrongly, accused of adultery. The chief, whose neglect of her had been matter of common talk and reprobation in the tribe, sent his executioner and killed her and her children. Immediately after that, he sent a messenger to inform us that he had married another wife—the twenty-sixth. When Mr Koyi remonstrated with him and said he thought he would be afraid to increase his troubles in that way, he laughingly replied, " I do it for peace ; this sets them on each other and they leave me alone."

Mombera had a dual character. He was at his best in the early part of the day, before he became intoxicated, and so by sun-rise people with cases to be judged went to see him. Then his affability and generous behaviour were pleasant to see, but toward afternoon when the beer he continually sipped began to act, his civility was at an end for the day and he was foul-mouthed and quarrelsome. When he was sober he delighted to play with his children, and manifested a very pleasing interest in them and their mothers, but when drunk he drove them from his presence with obscene curses. He had a great interest in old people, of whom he had always a number living in huts within the seraglio. He treated them with respect and provided for them from his own table. If he was shown anything new and strange he would generally have it shown to the old people, and while they knelt before him in due respect, one could notice with pleasure their trustful attitude and how he would heartily respond to any observation of wonder they might express. On one occasion he sent for my wife's sewing-machine that the old people in his village, who were unable to walk over to the station, might see it at work before they died. He said they would have to report to the ancestral spirits how many new and wonderful things had now become known to the people. When I went to exhibit its working, from some cause or other it could not be got to sew at all. In vain I tried to put it right, and Mombera, who had sat looking on with unusual patience for some time, unceremoniously rose and walked away, saying, "You need not try. You told your wife where you were going." As a polygamist ruler with many strings to hold in his hand, he believed that success is impossible if the wives are taken into confidence, and he supposed the same of us.

I have been a witness of some of the sweetest of life's incidents in the behaviour of Mombera to children and old people, just as at other times he has exhibited some of the darkest phases of heathen corruptness. But he was neither cruel nor bloodthirsty as many chiefs of the Zulu tribes have been. He discountenanced the poison ordeal which was adopted from the Tonga slaves, believing rather in their own trial by boiling water, which at most only maimed the person and did not destroy life as the *muave* did. He was considered to be " too soft " by the more degraded and fiery dispositions, and had no delight in condemning to death. Only two instances of the death-penalty being inflicted by Mombera came under my own observation, during all the years I lived under him. In one case he caused a man to be put to death for cattle-stealing, after having before pardoned him for the same offence. He hanged him from a tree near our house as a warning to those who about that time were stealing from us, and the body hung for three days before the white ants ate the rope and let the hyenas get it. The other case was where a member of the royal family killed a slave, who had run away from him and put himself under the protection of another master. Mombera by his action was esteemed more highly by the slaves, and he knew what would conciliate those who were the great majority of the tribe.

But despotic rule is often the only kind suitable among uncivilised people. Until the people are governed by higher principles than those common among " nature - peoples," a despotic ruler is a divine institution required to keep in check greater evils. I have been told by thoughtful old men that under Zongandaba, the father of Mombera, the Ngoni were purer, more truthful and more honest. Fornication, adultery, stealing and witchcraft were punished by death, whereas, under Mombera, capital punishment rarely followed these offences. The custom of the Tonga and Tumbuka of settling such cases by payments of goods had been adopted, and immorality had increased, while the respect shown by children to their parents and seniors had decreased.

CHAPTER VI

MEETING WITH THE HEADMEN

A WEEK after our visit to Mombera a messenger arrived to say, that next day we were requested to come and repeat our words to the head-men of the tribe. We had heard various rumours in the interval, which had caused us no little anxiety as to what would be the result of the meeting. It was said that I had come with many loads of calico, beads, brass wire, and all the many things the Ngoni desire, and at the meeting I was to enrich the people and make them great. Great was the excitement of the people over this piece of news. How such an idea came to them takes us back to the first meeting of Dr Laws with them, when the subject of war was referred to. Dr Laws had said that by obeying "the Book" and giving up war and plunder, they would become richer and greater than they were. The spiritual sense in which the statement was made was not perceived by the Ngoni, and from that day many were the theories expressed of how "the Book" was to bring riches and greatness to them. The native lives only for the present and could not be expected to see the force of such a statement, but it served to emphasise the special work we, unlike trading Arabs who were the only foreigners they had seen, had come to do. We were "the people of the Book" and not for trade. The Book was talked of, near and far, and became a source of wonder and enquiry, so that even from the start, while no systematic mission work was allowed, not a day passed on which some information was not given and seed sown, which, as we now view our work, has borne good fruit. It was no uncommon occurrence to see a group of strangers from a distance, at the house with the request to be shown the Book,—they had heard of it and wished to see it.

On the morning of the great council of *amaduna* we were in the chief's cattle kraal at eight o'clock, and the whole day till three o'clock in the afternoon was occupied in talking. The cattle-fold is the centre of every Ngoni village. At the royal kraal, where we met, it was a circular space about eighty yards in diameter fenced with young trees. Around it in ever widening circles the huts of the people were built. The gate was at the side nearest the river,

NGONI HEAD-MEN (EKWENDENI).

and at the opposite side was a smaller gate leading from the chief's quarters, which were fenced off from the houses of the ordinary people. In the centre of the cattle-fold there was one of the huge ant-hills which are so numerous throughout Ngoniland.

Soon after our arrival, troops of warriors fully armed marched in and took up their situations in the enclosure. There were eventually several hundreds present, but perfect order and quiet were observed. When all the warriors had assembled, the chief councillor, Ng'onomo, and the others came in. There were eleven present that day. Accompanying the councillors was a large number of men of inferior rank but possessing certain powers in the tribe. The councillors seated themselves in a semi-circle near to us. After the usual delay each saluted the Mission party, and then Mr Koyi rose to open the business. They were told I had come desiring to stay among them, and to teach them the Word of God, and to heal the sick. Several of the councillors spoke, and all were very warm in their expressions of welcome and readiness to give permission to my staying. All went smoothly until Ng'onomo got to his feet. He began by performing a war-dance, which, being accompanied by the war-shouts of the warriors

present, and as I could not understand its meaning, discomfited me not a little.   I was reassured
when I caught the broad smile on Sutherland's
face as he looked at me.

All the nice bits of native politeness and flattery
had been said, and Ng'onomo, bent on the one
question of war and conquest, desired to give the
meeting a more practical turn.   He finished his
war-dance, and after recapitulating the speeches
of the others, he plainly said that they were not
to give up war ; that they were accustomed from
their infancy to take the things of others and
could not see any reason why they should change
their habits.   He said, "The foundation of the
kingdom is the spear and shield.   God has given
you the Book and cloth, and has given to us the
shield and spear, and each must live in his own
way."   To emphasise this utterance, he again
danced.   We had adopted the plan of replying
to anything said when the speaker sat down.
Mr Koyi replied, saying that the Book was given
to all mankind, and that as we were all the
children of God it teaches us that we ought to
live in peace with each other.   Here I may say
that there is no word in Ngoni for "peace."
They now use an imported term,—their own
expression which comes nearest the idea being
"to visit one another."

No new question was raised at that time, but two crucial matters with the Ngoni in those days were brought up. They had been brought up when Dr Laws met the council, and for many a day constituted posers for us. One was the flight of the Tonga to Bandawe, and the other was their desire to have the exclusive right to the presence of the white men in the country. Mr James Stewart in 1879 visited Mombera, and wrote thus :—"The next day, Saturday, we reached Mombera ; but when I enquired for the chief, I was told he was 'not at home.' It was soon evident that he was either designedly absent, or that he simply denied himself. We saw only inferior head-men, who expressed dissatisfaction that we had not come to settle among them, and that they did not understand why we should visit other chiefs before doing so. I have no doubt that they were sincere in their desire to make friend-ship with us ; but an exclusive alliance would only suit them. We heard that they were tired of waiting for us, and intended now to take their own way, which, I fear, means war before long. They have lost both power and prestige within the last two years, and may now be resolving to regain both. I heard later that there are two parties in their council. Mombera and Chipatula and their head-men are desirous of peace and to invite us still to come among them, while Mtwaro and Mperembe wish to keep us at a distance, and to recover their power by force of arms."

Ng'onomo asked what I was to do to bring back their former slaves, the Tonga, who had revolted and carried away some of their wives and children, their war-songs, and their war-dances. So long, he said, as we would not restore these, so long must they war to bring them and all other surrounding tribes into subjection, and if I would not in a peaceful way bring back the Tonga people, they would do so by war or drive them into the Lake. It required not a little caution to answer this statement, so as to still the excitement of the crowd of people present by whom such words were applauded. I directed Mr Koyi to say that no doubt they had many questions in which they were deeply interested, but as I had only just come among them, it was scarcely fair to demand of me a means of settling them before I had become acquainted with them and had learned their language.

My remarks had the effect of drawing a very sensible speech from an old councillor. He said I was only now like a child, unable to speak or walk, and as they did not call upon their children to go out to seek strayed cattle, or give judgments in the affairs of the tribe, so they should not call on me to settle their great matters while I yet could not speak or walk. That statement turned the discussion into more favourable lines, and although the other question of leaving the Tonga and Bandawe and settling among the Ngoni exclusively was brought up, we were able to satisfy the people without exciting their jealousies, or agreeing to take sides with them against their runaway slaves. Ng'onomo afterwards returned to the war question, and endeavoured to show that their war raids on other people were not a bad thing. He said they were surrounded by people whom he called slaves, and that it was not their desire to kill them, but they endeavoured merely to chase them into the mountains, and when their food and flocks were secured, to say to them, "Come down now and let us all live together." It was conquest and not murder they pursued, as they could not bear the idea that any people should point the finger at them, and say, "X" (a click, expressive of contempt). He made an original proposal which was not less impossible for me to carry out. If we would agree to countenance one more raid on the people at the north end who were rich in cattle, and would pray to our God that they might be successful, they would, on their return, give us part of the spoil in cattle and wives, and would proclaim that the Book was to be accepted by the whole tribe. Here there was no place for parrying, and the reply was given emphatically enough that we were not the framers of the words in the Book, but merely the teachers charged to tell all men the words which were God's and binding on us as well as on them, and that when God said, "Thou shalt not steal," "Thou shalt not kill," we had no power to change the command, and could not in any way countenance their wars. Then Ng'onomo asked if we would shut the Book and not pray against them if they went out. I said I had come to teach these words and could not but do so.

An interesting statement was made by one old man. He had evidently watched the life and character of Koyi and Sutherland, and considered its bearing on the practical things of daily life. He began by saying they were glad I was a doctor, and hoped I had medicine to make Mombera live long. He went on to speak of other medicine which he thought we possessed of which they had no knowledge. He said, " We see you white people are not afraid to go about all over the country, and you settle among different tribes and become the friends of all. How is that?

You have medicine (natives think everything is done by medicine as charms) for quieting people's hearts so that they do not kill you. We cannot do so. We are not even at peace among ourselves. We speak fair words to each other, but that is not how we feel. We have also noticed that your servants are 'biddable,' and when ordered to do anything at once do it. It is not so with ours. We tell a slave to do a thing, and he says, 'Yes, master, I have heard'; but he does not do it unless he chooses. We hope you will give us medicine to make our slaves obedient, and to quiet our enemies." A better opportunity there could not have been for giving them a little plain instruction, and for putting in a word for schools which had been proscribed since the Mission began. Koyi, whose speech was as clear and pointed as theirs, made good use of his opportunity. He told them we had no medicine in their sense, but the words of the Book were stronger than medicine when taken to heart. He quoted the golden rule, and said, "That's the medicine for quieting enemies everywhere, and was that which made all tribes the friends of the white men." Then as to making servants obedient, he said the Book had words for both servants and masters. It told servants to be obedient and honour their masters ; and masters to be kind to and patient with their servants, and give them their due in all things. He added that our servants were obedient and happy because they were being taught the Word of God, and because they were not our slaves, but were paid their wages regularly. He advised them to try it among theirs, and it would have the same happy results. Then he attacked once more the stubbornness of the people in refusing to allow schools. He said in doing so they were refusing the medicine which they were crying out for. As a native only could, he ridiculed them, and by happy and forcible illustrations made them hesitate in the position they held in refusing to allow schools. He said, "You are like a sick man in distress, who sees others being cured and cries for the same medicine, but refuses it when offered." One replied by saying, "If we give you our children to teach, your words will steal their hearts; they will grow up cowards, and refuse to fight for us when we are old; and knowing more than we do, they will despise us." That was met by saying that the Book had a command for children which they must allow to be good, viz., "Honour thy father and thy mother." They would not be taught anything wrong, for all men are taught to fear God and honour the King. The school question was not discussed further; but no doubt some good was done, and the solution hastened by what had passed, although it was, as we shall see, two years after this ere liberty was given to open schools.

One other point it was necessary to refer to, as only the district immediately under Mombera was open to the Mission, so I requested leave to go about the country, as my desire was to help all. The districts of Mtwaro, Mperembe, and Maurau, brothers of the chief, were closed to us, not more by the hostility of these sub-chiefs, than by the jealousy of Mombera and his advisers, who desired to have the white men all to themselves, no doubt in view of the riches which were expected to come through them.

I was advised to stay with the others, as all were not favourable to our presence in the country; and while we would be guarded if in their midst, they could not tell what might happen if we went beyond Mombera's own district into that of any of his brothers. This was not satisfactory, and as it was probably from jealousy, we pushed for liberty to go about. It was denied by the councillors, who repeated their reasons.

It was, however, clear in all that was said, that the real object of our presence among them was made manifest.   However mistaken their ideas were as to the teaching of the Book, we were understood to be men with a message to be received, and they were honest enough to say they did not want it. No advance on previous liberties was made, but our position as neither wishing to bear rule over them nor to work for their overthrow, but to teach the Word of God, was made plain once more.

Then came the not very agreeable business of presenting the gift which we had taken for the councillors. There was considerable excitement visible generally, as each was presented with twelve yards of red cloth, a kind much valued by the head-men. As each had his portion presented to him there was an ominous silence for a time, and then a burst of derisive laughter. Some turned it over on the ground as if afraid to handle it. Some got up and measured it. One man took his and flung it among the crowd of warriors. One came over and said he did not want cloth. One only had the grace to thank me. They were reminded that we could not attempt to enrich them with goods, but had merely, according to their custom, brought " something in our hand " as a visible token of the friendship our hearts desired. One replied saying they saw we were not bent on enriching them, but it was good to remember that they had great hunger for various kinds of cloth and beads, and another day perhaps they would receive more. If I had come among them expecting the grace and politeness of civilization, instead of their proud indifference and sovereign contempt for the offering of friendship, my feelings would have suffered more than they did, but I was heartily glad when they rose up to go, and that the wild rumours of their expectations which we had heard for some days, found no more pronounced substantiation than their contemptuous treatment of what I thought was a sufficient gift for the purpose in view. The armed warriors, who appeared to have come as the bodyguard of the head-men, quietly filed out of the kraal and we were left alone.

Mombera was not present, and the councillors went to his hut to report to him the matters which had been talked over. Mr Koyi was called, and it seems the chief had enquired the reason why war dancing had been engaged in. He was angry at Ng'onomo and told him that the object of the gathering was not to discuss tribal matters with me, but to hear what I had to say. After a little the rest of us were called into the chief's hut, where Ng'onomo and some of the other councillors were being regaled with beef and beer. The stiffness and formalities of the kraal meeting were absent, and no disappointment was visible. Mombera delivered a long speech bidding me welcome among them, and expressing joy that I was skilled in medicine. He himself was often sick, he said, and doubtless I had noticed that there were few old men present that day, the reason being that they were all dead, and if I could give them long life it would be good. He did not say how many never reached old age because they were killed in battle. If there were any doubts as to the full security of our position in the tribe, they were accentuated when Mombera repeated the warning of the councillors, that I should settle along with the others and not go into other districts. No doubt there was some desire to have exclusive possession of the white men, but it was noteworthy that although word had been sent to all the sub-chiefs to come to the palaver none had come, and none of their head-men were present.

With too great eagerness, perhaps, I pressed for permission to visit his brother, Mtwaro, at Ekwendeni, saying my desire was to become acquainted with all in the tribe and be of use to all. He and Mtwaro were not on friendly terms at that time, but as Mtwaro was heir-apparent it seemed advisable for the permanence of our work, in the event of Mombera's death, to become known to Mtwaro and his head-men. Not since 1879, when Mr John Moir visited Mtwaro and had opened the way for others by friendly dealings with him, had anyone communicated with that sub-chief, and he had only once visited the Mission station. His armies were known to be out towards the Lake very frequently, and we all thought an attempt should be made to gain Mtwaro's influence as Mombera's had been gained.

After my statement had been interpreted to Mombera and he had consulted with some of those in the hut, he gave permission to visit Mtwaro and was thanked. He seemed to think that that would soften my heart, and so he plied his begging and his demands for cloth, beads, brass wire, big guns, little guns, gunpowder, dogs, bulls to improve his breed of cattle, needles, thread, and, above all, an iron box, with lock and key, in which to keep his valuables, which he said his wives and his councillors were in the habit of stealing. He said he would come over to see me when I could give him these things. It was hard to take all in good part and be at ease under his gaze over the beer-pot, and gracefully excuse our non-compliance with his overwhelming demands. Nothing but a desire to be a means of blessing to such a chief and tribe, would prove an inducement to live the life and experience which may be said to have begun that day. Forgetting the things not agreeable to flesh and blood, we soon after took our departure, feeling that some advance had been made in the work which we had come to take part in.

It was one advantage having to deal with a council rather than a single individual, and be continually subject to his capricious mind. As the Ngoni had a settled council who were not without dignity and caution in their deliberations, it was evident they had reciprocated our words as far as they could, as, not being over-anxious to allow us all we asked, they were prepared to make good all they allowed. The occasion was very similar to that on which Augustine came to Ethelbert as the first papal missionary to Britain. When he sent word on landing that "he had come with the best of all messages, and that if he would accept it he would ensure for himself an everlasting kingdom," Ethelbert would not commit himself, but answered with caution. When at last a meeting was convened, and Augustine "had preached to him the Word of life," as Bede says, Ethelbert replied, "Fair words and promises are these; but seeing they are new and doubtful, I cannot give in to them,

and give up what I and all the English race have
so long observed." But unlike Augustine, who
was accorded the privilege of bringing any one
of the people over to the new faith, we were told
that the chief and council would first have to be
taught, and if they considered our message safe,
they would give us full liberty to teach the people.

It may here be noted how different has been
the introduction of the Mission to all the other
peoples in Livingstonia. In all the other districts
the missionaries were hailed as the friends and
protectors of the people. All were subject to
stronger tribes, by whom they were constantly
harried, or were trying to maintain an independent existence surrounded by their enemies;
hence they gladly welcomed the missionary,
hoping that his presence would prove their
safety from their enemies. In no single case
did they welcome him on account of his message; and the trouble in those early days was
that he was pestered for medicine, guns and
powder to kill their enemies. The Missions in
those districts had the preparatory work to do
in making the people understand the reason for
their presence, just as we had of another kind in
Ngoniland. Through the faithful testimony of
Messrs Koyi and Sutherland, the Ngoni had by
the time of my arrival come to understand clearly
what our message really was. They needed not

our protection from their enemies, as they were masters of the country for many miles around; and, indeed, their pride would not have allowed them to think that in any way a white man or two could be of any profit to them. They knew our teaching would strike at their sins of uncleanness, lying, war, murder and stealing, and they were, unlike the so-called deceitful, vaccillating African, at least honest in their treatment of our words. There was great good in having got their ear so far; and even distinct refusal was far better than ready compliance, to be as readily retracted when occasion arose. It is far better to have to deal with an opposing council of head-men with power than with a chief himself, even although he agrees at the time.

If before leaving home I received one bit of advice more often than any other from Dr Laws, who had experience, along with Mr James Stewart and Mr Koyi, of the dangerous and trying work of gaining an opening among the Ngoni, it was that I should proceed gently and push nothing beyond what was a wise point. On such occasions as the meeting referred to, the judgment and caution of Mr Koyi were invaluable, and he was of opinion that we should not endanger our position with Mombera at that stage, while not sure that we would be received by Mtwaro. We sent a reply that we had no desire to act contrary to the chief's wishes in the matter, and that until he could send someone to introduce us to his brother, we would refrain from going. It must be remembered that we were merely in the country on sufferance at that time. We did not even own the site of our house, and were not by any means assured of a permanent residence among them, so that we would not have been acting wisely had we been more anxious to assert our independence, than to improve the, as yet, slight hold we had on Mombera and his councillors. There are three special qualifications necessary in every missionary, viz., grace, gumption, and go. Prayer and the exercise of it will ensure the first; where one may get the second, I know not, but the want of it is accountable for more failures in the foreign field than anything else; and the third, although invaluable, can only be right as the outcome of the former. To spend years among the Ngoni and be denied many liberties may, indeed, be an undignified position for a free-born Briton; but mere questions of dignity ought not to trouble the slaves of Christ in the work to which they have been called. Little by little, as we shall see, our position was improved among the Ngoni, and the years of apparent unfruitfulness were necessary preparation for the intelligent acceptance of the Gospel.

CHAPTER VII

ABOUT the time when I was beginning to realise how actual mission work differed from the romantic ideas of it too commonly entertained at home, and overcharged with which many enter the field, a notable missionary—A. M. Mackay—far away in Uganda was writing these words :—" Current ideas at home as to mission work are, I fear, very different ; but I have not heard of any part of Africa, east or west, where the native bearing to the Missions is different from what it is in this neighbourhood. It is a system of *beggary* from beginning to end, and too often of suspicion, and more or less hostility too. Only when these first adverse stages are passed can we expect to do any real good. Disarming suspicion and securing friendship are a slow process, but an absolutely necessary one. They are most wearisome and trying to the faith and temper of those engaged in the task, while they yield no returns to show in mission reports ; yet on their success depends the future of our work. Hereabout we are so far from the *reaping* stage, that we can scarcely be said to be *sowing*. We are merely clearing the ground, and cutting down the natural growth of suspicion and jealousy, and clearing out the hard stones of ignorance and superstition. Only after the ground is thus in some measure prepared and broken up, can we cast in the seed with hope of a harvest in God's good time."

These are words of truth and soberness, as every real worker can testify from his own experience. At this time, being unable to move about among the villages with any degree of freedom, we were often compelled to pass the time on the station, and were assailed by overbearing and impudent men and women, clamouring for whatever they saw with us which they coveted. To say we were annoyed is to use a mild term for our experience. From morning till night the house was beset by natives begging. They allowed us no privacy, and our rooms were darkened by a crowd pressing round the windows and flattening their noses against the panes. If one ventured out his steps were dogged by a clamouring mob. Any attempt to divert their attention from begging by showing pictures, explaining the work-

ing of apparatus, or manufacture of articles, was
treated with indifference.   Time was of no value
to them, and so for many a long day the vicinity
of our house was the meeting-place of all who
sought diversion through watching the white man,
or begging for the clothes off his back.   Men who
could have been well clothed appeared in rags,
which they took pains to show.   Others would
come in a nude state, hoping to appeal to us
thereby.   When they wanted cloth and beads
they complained of hunger, which they indicated
by drawing themselves in and simulating an
empty stomach.   If one offered them food they
disdainfully rejected it, and explained that their
hunger was for calico.   Their importunity and
arrogance were at times almost maddening, and
sometimes the only relief got was by shutting
up the house and going away to spend a few
hours on Njuyu mountain and leaving them
alone.   We could not reason them out of their
begging habits.   They could not entertain our
view of the disgraceful and undignified habit.
They would say in flattering terms, " We are
praising you by begging.   Do men beg from
people who are poor and mean ? "

But while the annoyance was great, their un-
reasonableness and selfishness made it well-nigh
impossible to bring any sort of benefit within

their reach. When we began to make bricks for housebuilding, and were thereby able to put some cloth in circulation among them, the work was repeatedly stopped by some head-man or combination of natives, who desired that they only should have the benefit of it. The very people who had been the friends of the Mission at first became our enemies, and did all in their power to compel us to submit to their demands to supply them with whatever they wanted. They had given up the spear and had been coming to our Sunday service, but as we would not enrich them with earthly possessions they turned against us, and reviled us for having cheated them, as they were now poorer than when they followed their own ways. Three brothers, Chisevi, Injomane and Baruke, the heads of the neighbouring villages, became openly hostile and threatened to go to Bandawe with war, because we would not pay them for being at peace with us. Injomane—the murderer of his own mother, cruel and treacherous—set out and attacked a village near Bandawe. On his return the war-party made a demonstration at the station, by engaging in war-dances, and speaking against the Mission and the " news." The effect of these war-parties going out was that we were left without mails and supplies at times, as the Tonga at Bandawe, on whom we had to depend for carriers, were afraid to venture on the road.

From the native point of view, those members of the Chipatula clan who had befriended the Mission, and had been the means of our gaining an entrance to the country, were right in attributing their position to their friendship for us. They were the sons of a once powerful chief who had lost his kingdom. They hoped that through the Mission they might regain their former position. They had heard and accepted Dr Laws's statement, that by serving God they would attain to greater riches than by using the spear. They did not apprehend the spiritual aspect of the case and gave expression to the only need they felt. Their expectations had been disappointed and they had, in befriending the Mission, become to a certain extent outcasts from the Ngoni who were all along opposed to the settlement of the Mission. They had not learned to work and now that their spears brought them nothing, they were indeed poorer in all that they valued. It was often a trying situation to meet their attacks and to quiet their feelings, and in it all we saw how not the words of man but the Divine Spirit, can reveal to men their spiritual state and make plain to them the Word of Life. It was peculiarly hard on William Koyi, when alone among them, to hear the Gospel accused in this way, and with a better intention than judgment he made presents to them to keep them quiet. He was discovering that it was an unsafe kind of peace which was thus produced, and when I arrived the whole question was discussed. We resolved that such a practice must be stopped.

As time went on matters did not improve. When our determination not to pay anyone for coming to hear the Word preached, or to give presents in answer to the demand of those who came to beg, became evident to them, they used other methods in trying to coerce us. Our cattle were stolen from the herds when feeding, or from the fold at night, and we were never able to detect the thief. Trees brought in for firewood or housebuilding disappeared ; clothing hung out to dry was stolen, and our fields and gardens cleared of produce. As we were living among them on sufferance, there was no healthy sentiment to which we could appeal when wrong was done to us. If we could not detain the thief in the very act there was no case. During the rainy season we frequently suffered from cattle-stealing. On a night when rain was falling heavily, the fold would be entered and the best beast taken out and driven far away before morning, the heavy rain obliterating all trace of

the route taken.   The time of service or prayer-
meeting was chosen for entering the corn-field
and garden, and stripping them of our food
supply.   It would have been very easy at any
time to produce a rupture between us and the
natives by a want of forbearance on our part,
and yet there were circumstances at times, in
which it was impossible not to defend our property though not by force of arms.   On their
part they made war demonstrations on the
slightest occasion.   The cattle-herd may have
allowed our cattle to stray into a native's
garden, and he and his friends would come to
the station armed and perform a war-dance
as a preliminary to opening the case.   Nothing
was so effectual in overpowering them on such
occasions as quietly to allow them to dance till
they were satisfied, and then calmly say "Good
morning."   When the season for beer-feasts came
round we had to live through much that was
exceedingly trying to flesh and blood, and could
only be endured for the Lord's sake.   The beer,
which was brewed from a kind of millet, was
considered "ripe" after so many hours' fermen-
tation, and in order to annoy us it was frequently
made so as to mature on Sabbath.   Then early
in the morning the guns would be fired or a
horn blown to inaugurate what would be a day's

A VILLAGE SCHOOL.

debauch, and the people congregated for the orgie. As the hours wore on and the drunken natives began to dance and sing, the sacred day was filled by unhallowed sounds, while towards evening what had begun as friendly song and repartee, ended often in fighting and bloodshed. Our quiet was not only broken by these sounds from the villages, but sometimes a band of drunken youths, or men and women, would come to the service or to our door and assail us with foul song and epithet, or engage menacingly in war-dances.

In July 1885 an attempt was made by Injomane (before mentioned) to frighten us into resiling from our position on the question of presents, and the issue of which considerably strengthened our hands. A party of Tonga had come up from Bandawe with letters and goods. When they had gone a few miles on their return journey, Injomane and a party of his young men attacked them. They were robbed of all their clothing and their weapons, and some of them wounded. Chisevi, a brother of Injomane, came to the station and informed us of the threatened attack, hinting that while he had a good heart to us himself, he had, for the sake of his position, to appear at times as our enemy, and that we would no doubt see how he esteemed us and reward him for informing us. Before we had time to act for the protection of our Tonga carriers, one of them who had escaped without wound returned to give us information. The others, wounded and robbed, escaped into the bush, not daring to come back through the villages in a nude state. We considered that the case should be taken to the chief, in order that we might see of what value were the words of the chief and councillors in protecting us. Mr Koyi and I thereupon went to Mombera and made complaint, pointing out that protection to us must mean also protection to any in our service. Mombera, with his natural shrewdness, asked us why those who had brought us into the country had now turned against us. We said that they were harassing us because we would not satisfy their demands for cloth and beads. He was very angry and called the Chipatulas " rats," saying that it was only our presence that preserved them from the attack of his army. He desired to send an army over to punish them, but we proposed that he should send a councillor to make an investigation and call the people together to inform them that we must be protected.

Ng'onomo, his prime minister, being the councillor for the district in which we lived, was sent to hold a court. All the villagers were called up, and although Injomane and Chisevi (who had informed us) denied all knowledge of the affair, after a whole day's talk, Ng'onomo decided that Injomane had done wrong and that the cloth and spears should be returned. We were asked if the punishment was full enough, and we had opportunity of expressing our regret that the people in whose interests we had come should not admit us to their friendship, and permit us to carry on our work for their good. After warning the people against annoying us, Ng'onomo declared the *indaba* at an end. An ox was killed, and the judge, prosecutor, and defendants all feasted together in amity. The Chipatulas had feared other treatment, as they had sent away all their herds and goods, so that they had another exhibition of our forbearance and desire to do them good.

If we had been asked by carping critics at this time, "What are the results of your work?" we could not have pointed to a single convert, although the Mission had been already three years in the district. To all appearance it was a failure. From the chief and the councillors we had stolid indifference, and direct veto against educating the children, or moving about to preach the Gospel; and from many of our near neighbours we were receiving marks of base ingratitude and opposition. But was no work being done and no good being accomplished? Of stated work there was not much. We were denied access to every village save two outside the area of Hoho, as the district in which we lived was called. On the station we were meeting daily with men and women, and youths and maidens, who were employed in housebuilding. To these we had opportunity of speaking about spiritual things. There were the boys in the house as servants who were collected for worship and oral instruction every day. A few young men outside began to take an interest in these services and attended. From them grew a stated service on the Sabbath, to which by and by others came, and although open preaching of the Word had been proscribed, we gradually came out more boldly and our service was tolerated, and in turn became an object of interest to others abroad. Only a few of the women came, and the men were fully armed.

The service was often very uproarious. The dogs snarled and fought with each other, and when this took place the "backers" of the different dogs whistled and encouraged them. Often audible remarks followed the reading of passages or parts of the address. Sometimes a man would get up and declare that it was all lies, and demand cloth as they had heard enough of the Gospel. Some came out of curiosity ; others came having the impression that we gave cloth to all who attended ; and sometimes spies were sent by the chief's councillors to see and report what was done. This was known to us for some time, but we did not think any evil would come of it, until the rumour got abroad that we were inciting the slaves to revolt against their masters. Mr Koyi had the burden of anxiety for he heard all that was being said, and was always either the preacher or interpreter, as I had not then acquired the language. The rumour arose from the Tumbuka slaves having begun to attend the meetings, and afterwards discussing the teaching of the ten commandments in the villages. Their masters began to be suspicious, and for a time we feared that our service would be stopped. "The common people heard us gladly," and were realising that in the Gospel there were hopes unfolded for them which found a response in their hearts. We were called to account by the councillors, but were able to satisfy them as to what was said and done, protesting that we had no desire to interfere in their tribal relationships or to upset the authority of the chief.

As young men we were used in exercising an

influence on the young men very particularly, and gradually gathered round us a band of half a dozen, who began to speak in defence of our work. They even met together for prayer and singing of hymns, and were in consequence marked out for persecution. They were called " bricks," in derision, as they worked with us and favoured us. They were often set upon by others, and had many a hard day, while yet but imperfectly taught in the Word. But it was the beginning of fruit, and came to brighten our labours. To show how the changed behaviour of those lads led them into trouble, the following instance is given. The child of one of them was ill. Although the grandfather was a native doctor, the father called me to attend his boy. He was suffering from croup. It being the custom for the father not to appear in the presence of his mother-in-law, he could not enter the hut where she was. After treating the child I went away, but on my next visit I could not find my patient. It had been carried out into a maize field. I saw the poor thing struggling for breath, and soon after it died. The " smelling-out " doctor was called to discover the cause of death. He decided that the spirits were angry, and wanted to punish the father for forsaking the beliefs of the old people and listening to our preaching. He had also been neglecting

to offer sacrifices to the ancestral spirits. So
strong is their faith in their doctors that all this
was believed, and our young disciple had to suffer
persecution.

While the direct evangelistic work was circumscribed, there was practically no limit to the
medical work which I carried on in the district
ruled by Mombera. At first people came in
crowds. Those who were sick expected to be
healed immediately, and those who were not
sick expected medicine to keep them well. Many
cases of a very trivial nature were treated, but
there was a value in the work apart from the
relief given to the individual. For instance, if
a slave were sick and unable to work, no care
was taken of him. Such were sought out, and
often a master had a useful servant restored to
his service. He put a value on this, and was
favourably impressed with this part of our work.
It was easy to get a hearing from such as he on
the other aspects of our work afterwards. A poor
woman, left to die as an evil-doer if she failed in
her "hour of nature's sorrow," when saved, together with her infant, by treatment of the proper
kind, would thenceforth be well disposed towards
us and our work. A wife represented so many
cattle, and her husband would appreciate the
benefit of our work and be our friend. Little

children, relieved from pain and sickness, understood the practical nature of the work, and would always respond to our words. In such ways, up and down the country, the work was quietly and surely influencing the people, and while there was yet nothing to tabulate for reports, the future harvest was being insured.

Many things compelled the people to talk of us and our work, and it was plain that while there was no sign of liberty being given to teach the children and preach throughout the tribe, the feeling among the people that we were not being sufficiently trusted was gaining ground. We took advantage of any opportunity to renew our application to be allowed to open schools. Sometimes that led to their discussing the question, and at other times it led to threats to withdraw all permission to preach. We began to be more respected, as those who had received benefit were bold to declare it, but we did not seem to have made any impression on the chief and councillors. They continued to declare that they would never receive the Word of God, while the common people said that until the heads of the tribe did so they could not. The reason why the head-men would not countenance our work was no doubt because they knew that the result of it would be to overthrow their power over the slaves, and to crush the war spirit in their children ; also, because they were in the hands of the witch-doctors, whom they trusted to the utmost as the only channel of communication with the ancestral spirits. Those witch-doctors were against us as they saw their craft to be in danger.

One of the greatest effects of the medical mission work was that, by it, the empiricism of the native doctors was overthrown, and the common people, ignorant and superstitious, were rescued from the bondage of their shrewd but deceitful incantations. Native doctors fail in diagnosis more than in power to heal. Yet in the presence of the majority of diseases they are helpless, and in that case they fall back on the professed will of the spirits that the patient is to die.

Towards the end of this year (1885), having received encouragement from a sister of the chief who was head of a village called Chinyera, about five miles from the station, we built a round hut there and Mr Williams went to live in it. When this came to the chief's ears he sent for us, and asked if the country had been given over to us that we had begun to occupy it. We referred him to his sister who had invited us, and we heard no more of it although it led to increased bitterness among the councillors. We had thus actually, without formal liberty, opened our first sub-station and widened the area of our influence. Mr Williams conducted a small service in his hut, and Mr Koyi remained with me at Njuyu doing the same work. But during all those months we were the subject of continual discussion among the people. Sometimes a councillor would spend half a day on the station speaking on things in general and evidently having some errand which he was unwilling to reveal. In going away he would ask, "How long are you going to stay among us seeing we are refusing your message?" What to make of us or what to do with us, was evidently a problem which they could not solve. They were no doubt irritated by hearing of the prosperity of their former slaves, the Tonga, under the Mission at Bandawe. We were considered to be standing in the way of their compelling their return to bondage, and over and over again disquieting news of what they were saying and plotting reached us. It was a common occurrence for a section of the army to be called up for review and to get secret orders. Not only our own position, but the position of our brethren at Bandawe gave us anxiety on such occasions. Sometimes the Chipatulas would suddenly show us great kindness, and inform us that Mombera's army was to attack them and

us.   On several occasions the neighbours set
watch at night and made preparations against
being attacked.   Our friends at Bandawe had
anxious times too, on our account.   Once the
letter-carriers coming up were informed of the
expected attack at a village on the outskirts of
the tribe, and in fear returned to Bandawe without coming near us, and our friends were left in
doubt as to our safety.

It was in the end of 1885 that the first expressed evidence was given that the Gospel was
winning its way into any heart.   At the close of
the boys' meeting on a Sunday evening, Mr Koyi
had the joy of hearing from Mawalera, who had
been in his employment, that he wanted to pray
to God.   After he had poured out his heart in
broken accents others joined in the exercise, asking that God would teach them to pray, and give
them hearts to love and fear Him.

Notwithstanding this new joy and the strength
it brought us, we were soon in deep anxiety on
account of the persecution which was levelled at
the youths who had begun to confess Christ
among their fellows.   In Matabeleland no sooner
did a native confess Christ than the chief ordered
his execution, and at that time we were reading
about the burning of converts at Uganda.   We
told our young friends these things and asked

them to count the cost. They were not borne up by any unusual emotion, but they expressed themselves prepared to witness for Christ. The occasion was seized by Chisevi, one of the Chipatula clan (our neighbours already referred to) as suitable for our overthrow on account of our refusal to enrich them. He went secretly to Mombera and informed him of what had taken place. Mombera showed his aversion to the informer and his great friendship for us, by receiving the report without a word. Afterwards on a visit to the station he referred to it, and the conduct of the boys was defended by Mr Koyi, and beyond the persecution which the boys met with, no evil resulted as we feared might have been the case at the time.

The year had seen our hearts bowed down in sorrow by the death of our brother Sutherland, whose life and work are referred to at length in another chapter. We had now at its close the joy of seeing the ingathering of the first-fruits of the work, in which he was for a time associated with Messrs Koyi, Williams, and myself, before another cloud was cast over us by the death of Mr George Rollo, who had just come from Scotland to begin work at Bandawe. He arrived on Mission duty at Njuyu on December 21st, suffering from fever, which, with one day's intermission, continued till the 28th when he died. As marking the attitude of the people towards us, when Mombera came to know of his illness he requested us to take him away lest he should die in their country, and when he died we were accused of bringing him to the station to die, in order to involve them in trouble which they ignorantly feared might come to them on account of the death. They proposed that we should take the body away and bury it at Bandawe, but eventually a grave was opened near the station, and the object-lesson of a Christian burial given to the natives, who gathered together at a distance and looked on.

CHAPTER VIII

THE RAIN QUESTION AND ITS RESULTS

MY journal for 1886 opens with this entry,—
"*Jan.* 3. In bed three days with fever."
Notwithstanding the unwearied assistance rendered by Messrs Koyi and McCallum during the strain of nursing Mr Rollo throughout the week of his fatal illness, I was worn out, and I had a sharp attack of fever myself, the usual result of over-anxiety and fatigue. Thus began the year that was destined to be one of sorrow and of joy for those at Njuyu, and of the triumph of the Gospel among the Ngoni.

About this time the station went by the name of "Ekusinda-nyeriweni," a term which cannot be translated in polite language. The name was given by Mombera, and although it was accepted as a bad one by the people, he did not mean it thus. It arose out of the frequent complaints which people took to him of our supposed evil powers. We were accused of all the family disasters; the non-success in battle; the death of cattle, and the running away of slaves, or whatever evil came among the people. Mombera, who as may be seen was a believer in us, became so irritated at their numerous charges that he said, " The people are surely comfortable now that they have got a " Ekusinda-nyeriweni." The rebuke was levelled at them and not at us, but the name stuck to us for a long time, until we got a new name from the councillors which will be mentioned further on.

As we were treated with suspicion it was doubtful what effect our having brought a horse into the country would have. Messrs McCallum and Rollo had come up with a horse which was the first that the natives had ever seen. Before they actually arrived the commotion over this strange animal which they were riding was very great; and wild and absurd stories as to its appearance and behaviour went round the country. It was said to have only one eye, which when turned on one felled him to the ground; it was as tall as the highest tree; its feet crushed houses and people; its bounding step enabled it to jump over mountains and rivers; it had a tail which moved continually and smote people to the earth. Such were the wild impressions which this horse made on the ignorant people, who had only heard of it from others as it approached the villages. Although, when it was seen, the people became intelligently interested in it, we were in difficulties as to pasturing the animal. Complaints were lodged that we allowed it to go near their villages, so that child-bearing women could not come out, the belief being that the strange animal would lead to the birth of monsters. We were even advised not to allow it to come near the herds of cattle for the same reason. But gradually their fears subsided, and instead of being regarded as an evil thing, the people came long distances to see the wonderful animal.

Although, as mentioned in the previous chapter, Chisevi's informing the chief about the youths who were coming to us and praying, had not led to an attack upon them as was at one time threatened, the persecution they had to suffer was so great that when they desired to be taught to read they were afraid to come by day, and so they came under cover of night. At first the three sons of Kalengo, a witch-doctor, who was our nearest neighbour, came. Their names were Chitezi, Mawalera and Makara. We spent several hours together every evening, and they made rapid progress in reading and writing. They were also instructed more fully in Christian truth. The devotion of these youths was most

marked, and as we watched their minds opening
under instruction, and their hearts and con-
sciences coming under the deepening influence
of God's Spirit, we felt stronger and more hopeful
in our work which was so liable to be stopped by
the superstitious clamourings of the people. But
Mombera was no doubt cognisant of all that
went on, and it was noticeable that he began to
look upon Chitezi as our man. He being the
only one of the three who was married could be
accorded the place of a man, and so Mombera
and we had him as a common messenger on
nearly all occasions of communications passing
between us. It was a gratification to see the
respect which Mombera paid to Chitezi even
after he had thus cast in his lot with us. Chitezi's father was much respected, and Chitezi
himself had but lately been distinguished in
several fights and had received some special
marks of the chief's appreciation of his courage
and prowess. Yet his turning to us, while
against the expressed desire of the council, did
not lead to Mombera (who knew all) turning
against him or us at that time. Many other
things were known which betokened that the
mind of the chief as an individual leaned towards
our work, however much he spoke in public to
the contrary. He had a dual nature,—on the

one hand he was set to curse the work on behalf of his advisers, while for himself, he was, consciously or unconsciously, serving God's purposes and helping it on in many ways. On many occasions we had to thank God for the presence of even the heathen Mombera on the throne.

When these youths met for prayer it was very touching to hear them plead for the enlightenment of their father the witch-doctor, and for their friends, their chief, and head-men. They had, as youths, understood the worship of the ancestral spirits, and appreciated the position of prayer in the new life. It was a powerful inspiration in this exercise when they apprehended God as Father, ruling, guiding, and sustaining the world, and the need and opportunity of coming to Him in calm as well as in storm, in prosperity as well as in adversity; because in the ancestral worship they did not require to think of the spirits except in case of sickness, famine or drought. It was very interesting to watch the development of their minds under the influence of the truth of Scripture, and how the mind, accustomed to slavery and the relative positions of master and slave, chief and vassal, which the system entailed, naturally assumed the same forms under the spiritual kingdom. While they acknowledged God as Father, under fuller instruction in the Gospel of Jesus Christ, the idea of Him as "Great owner of power," "Owner," "Chief," was what came naturally to them, and by these terms they usually addressed Him.

It was an unspeakable relief to us when we had actual members of the tribe to consult with on the spiritual phases of our work, who were able to read the meaning of the many disquieting things which occurred around us in the behaviour of the people. We felt there was a bond between us not born of earth or earthly power, and in the exercise of prayer we could agree as touching anything we asked of the Father, and even then we could count six,—a European, two Kafirs, and three Ngoni, — as with one heart and mind desiring the coming of Christ's Kingdom. There have been many joyous seasons in my experience of the work since, but none have left such an impression on my mind as those of the time we write of. Our anxieties were unceasing; our position in the tribe insecure; our efforts all but fruitless among the great mass of heathen; our bodies frequently racked by fever and sickness; we had but occasional communications from home; but after dark the three Ngoni youths came to join with us in prayer for the work. They had staked

their safety and their position in the tribe in
accepting Christ.   They had their temptations
and their fears to relate, and we could hold common converse on the outcome of events, and
encourage one another in our trying circumstances.   Hallowed, indeed, were those hours in
the stillness of night, and as we knew not what
a day would bring forth, but continued in prayer,
we are now able to look back and see how prayer
was answered, and in that little sanctuary in the
dear old house in Njuyu the faith of that little
company has brought, by the mercy and over-
ruling hand of God, a rich return.

There is one phase of Mission life and work
which is not often written upon, but which ought
to be mentioned.   At home men and women are
called to volunteer for the mission field prepared
for sacrifice, and too often the idea of a sacrifice
which must be made is the one most prominent
at such times.   It is a false position in which to
put the work.   Why not keep before the mind
the advantages to one's spiritual life in the work?
I am not the only one who has felt that the Gospels and Epistles, as well as the Old Testament
Scriptures, have a fresher interest and newer
meaning to us when we are teaching the simple
minds of the heathen; and that the exercises of
prayer and faith in the circumstances of the new

life are more real and refreshing. One learns the simplicity and reality of trust in God when he hears a native, who may only have a few ideas or facts of divine truth, pouring out his heart to God in earnest request, and waiting with expectancy the answer to his prayer. Does God hear prayer? Our three lads had learned as much of the truth as enabled them to believe and ask, and one of many special objects prayed for, may be stated as it occurred and confirmed their and our faith in the presence and power of God, and His care of the work.

The occasion was when the increasing wealth and number of his wives compelled the chief to make choice of an additional royal residence. He had seven or eight royal kraals, and now he was to found another. It must be remembered that all this time the whole tribe, save the three youths whom we were instructing, were given to war and raiding other tribes. It was the custom in connection with the founding of a new kraal, to call up the army and make a raid on some tribe, setting the young warriors belonging to the village chosen as a royal residence in the forefront of the battle, in order to test their valour and ability to protect their chief in his new kraal. From what we were told we knew such occasions to be times of great excitement in the country, and the war following a very bloody one. The young bloods had to "wash their spears in blood," and it was their ambition to have an important battle to prove their valour. Our boys were greatly distressed—especially Chitezi, who would have to take his place in the Hoho regiment under the Chipatulas who were our oppressors. The turmoil went on for some days, and we heard that the army was to be despatched to attack the Tonga on the Lake shore around Bandawe. On the day when the royal entrance into the new village was to be made, we hoped that some opportunity might be had for Mr Koyi to speak to Mombera to advise him against sending out the army, and we prayed that Mombera might be restrained from ordering war. We heard that Mombera had been debarred from entering his village by the armed youths, who demanded of him an order to go out and "wash their spears in blood," that the chief had refused and was sitting outside determined to occupy the village without giving a pledge to order out the army. The armed escort that accompanied the chief to the new residence were in an excited state, and were threatening to fight the others who were resisting his entrance. As darkness began to fall we could see bodies of men rushing hither and thither among the villages beyond the river, and we feared that it would end in disaster. We decided that Chitezi should go over to quietly watch the course of events, he having volunteered to do so, and that we should continue in prayer for the prevention of war. He returned about ten o'clock and reported that after a time of great excitement the chief was ultimately allowed to enter, and the warriors dispersed. We ended our day of prayer by acknowledging in praise the goodness of God. The event made an impression on the minds of all, and our faithful three had their faith strengthened.

If it should seem strange that a band of youths should so oppose their chief, it must be remembered that war overruled everything else. An armed party could steal cattle or anything it wanted with impunity, and I have heard Ng'onomo, when dancing, calling Mombera a coward because he did not order the army out. It was an understood thing, and would be done in order to give evidence of a man's readiness to serve his chief at any cost, and it was always accepted in that sense.

We were never long without some pressing trouble, and sometimes the anxiety was continued through many weeks. The anxious position no doubt frequently induced or accentuated the attacks of fever which all the members suffered from in those days, the attacks being more frequent and severe than those of later years.

The question of a famine in consequence of drought was agitating the minds of all in the tribe. A few showers fell in the November of the previous year (1885), and the people had planted their maize. It sprang up for a fortnight, and then, as the rains ceased until the 18th of January, the corn was burned up and the people began to be greatly excited. The usual period when rain may be expected is from about the end of November to the end of March, so that towards the middle of January, when the early sowing had been fruitless, and day after day the sun beat down from a cloudless sky and rendered cultivation impossible in the absence of rain, the excitement of the people, with famine staring them in the face, is not to be wondered at.

However irrational the native may be in his beliefs and practices he understands that there is no effect without a cause. In the worship of the ancestral spirits when they are supposed to cause evil by being displeased, the witch-doctor and the family or community recognise their responsibility, and possibly misconduct, towards the spirit of the house or the tribe. The practice of the witch-doctor is a fattening one, as he not only gets his fee but a good piece of the meat he recommends them to sacrifice to the displeased god. When we became "Ekusinda-nyeriweni" we expected that the witch-doctors, as well as the people generally, would hold us to be the cause of the drought. For some weeks we were left ominously alone by most people, and especially by those about the chief, but our faithful three managed to keep us informed of what was passing in their meetings about the cause of the drought. We were indeed blamed, and particularly as I had erected instruments in the garden to control the weather. These (meteorological instruments) I was known to consult morning and evening and to write in a book what I was doing. At this time, of course, a book was in their eyes nothing but an instrument of divination, and as will be seen, they believed that it told us what was in their minds. They spoke about "The Book," as the Bible was so often referred to by us, and they thought there was only one book.

We were not very anxious for a time. They were sacrificing cattle to their ancestral spirits—household and tribal—and although there was a general clamour about our being the cause of the drought we were not molested. But as the

drought continued and their sacrifices were un-
availing, more attention was paid to us and our
actions. Some people would stand at the hedge
looking into the garden, and discuss the probable
action of the meteorological instruments to which
they had seen me attending regularly. On one
occasion old Maumba, a councillor of the chief,
came to talk about the rain not coming and said,
"Why do you not give the rain? What does
your Book tell you is in our hearts about you
just now?"

At length a meeting of the doctors was called
to ascertain the cause of the drought. Till then
I hardly expected that the doctors had a good
word to say of us; but when, in answer to the
question whether we caused the rain to stop,
they made a united statement that we had nothing to do with it, we were greatly surprised and
pleased. The doctors were divided in their
opinion as to the cause of the drought. One
party made the cause out to be the strife between Mombera and Mtwaro his brother, as
the spirits were highly displeased therewith.
Another party said that the spirits were at war
among themselves, and the rain would come
when they finished. The third party said it was
true that the spirits were displeased, but not on
account of Mombera's quarrel with Mtwaro, but

Hora Station Scholars.

because the tribe had given no heed to the message which we had brought to them. He instanced what seems to be a fact, that one of their fathers, who died while they were at Tanganyika, and who had never seen a white man, told them that in the course of their wanderings they would meet with white men who would be their friends, and to whom they must listen and be friendly. They were thus neglecting the advice given them, and the spirits were angry. It did not, however, occur to them that to obey was better than sacrifice ; so they renewed their offerings to appease the spirits, and after waiting for a time they were still disappointed. Several talked to us in a quiet, suspicious way, as if insinuating that we had better send rain to save ourselves. At length several councillors and a large number of men came over to us from the chief's place. They came to ask us to pray to our God to send rain, as their own methods had entirely failed.

The councillor who spoke made an apology why they had not settled the question of schools. We asked him whether he had come to speak on that question or about the drought. He said it was the drought, so we said he did not need to introduce the subject of schools, as that had no connection with the drought, although we were glad to see that they still retained their sense that they had not treated the question as they ought to have done. We asked them if they came to us because they believed that we had the power of giving or withholding rain, and one of them replied, " We here to-day do not think so, but I cannot say that there are none who think so. We believe it is the spirits." We said, " Why do you come asking us to pray for rain when you do not believe in our God?" " Oh," he said, "just to see which is best." We asked if they would give up their own beliefs, and permit us to instruct them in the Word of God if they found that our God answered our prayers; but no one replied. We then for more than an hour preached the Word to them, explaining how their ignorance made them think God was only to be sought when our own efforts failed. We pointed out that they themselves believed that when the spirits caused any calamity or, as they thought, withheld rain, they knew there was a reason for it, and the doctors were called to find out the reason, and in all cases it was in themselves; and so, before they could expect rain, or what they wanted, they first offered sacrifices to satisfy the spirits; that we believed it was for something in us that God withheld his blessings, and so it was needful for us to repent ; and that this feeling of the necessity of repentance and of

a sacrifice proved that we were all similarly constituted, and that for us God had provided the
sacrifice in the person of His own Son. It was
a splendid opportunity for preaching and we had
close attention.

We sympathised with them, and said we would
make special prayer for rain in our meeting next
day. They wanted us to go over and pray in the
chief's cattle kraal ; but we refused, for the reason
that we wanted the people to come to our service
on the station, and did not wish the Bible to be
over at the chief's on such an occasion, because
they attached a superstitious importance to the
Book.   While we were engaged speaking, a war-
party appeared on the road some distance from
the house, and engaged in war-dances.   We did
not take any notice of it, although we knew it
was a signal of defiance to us for something ; and,
as we afterwards learned, it was held in readiness
so that if we had received them as they expected
by saying, " You have refused our word for these
years, why do you come now?" it would be called
up to dance in front of the house, that being the
only thing that the Ngoni can do when they are
nonplussed.   The councillors who were with us
were uneasy, very uneasy, when the party came
in sight, and no doubt felt relieved that we did
not run for our fire-arms, like the neighbouring

villagers, who were listening in the verandah, and who, on going home, found that their wives and children had fled up the hills behind the station.

We were never able to discover their real intention in coming with a regiment of armed men. It was only known to the councillors, and Mombera afterwards said he did not know either; but there may have been some idea of doing more than frightening us by it, because we saw another regiment making for Chinyera, where Mr Williams was at the time, and it remained in his vicinity for some time. Apparently some signal was made, and it returned to the chief's kraal soon after the deputation withdrew. Great was the excitement among the Hoho people around the station, and notwithstanding their conduct towards us, they now declared that our cause was theirs, and that as they had brought us into the country they would have to die with us, as that had been determined, they said, by the councillors, should we not be able to give rain. That night, as during the evening, armed men had been gathering at the chief's kraal, which was only a mile distant across the valley and in view of the station, so neither we nor the natives near us retired to rest. It was affirmed by all that we were to be attacked, and the natives set watchmen on all the ant-hills between us and the river. We did not so much fear an organised attack, as that some of the young bloods, excited by the war-dancing, might break out and fire the station, in the hope of really inducing war, and so " we made our prayer unto our God, and set a watch against them."

A touching word was spoken by old Kalengo, the father of our three adherents, who sat till far into the night with us at our house. He was a slave of the Ngoni from the Senga country, and had known the position of a slave in the tribe, till he became a witch-doctor. He feared the wild warriors who were collecting at the chief's place, and said, " Well, I'll go home to my own village now. If we hear the sound of war we will come to your house to die with you. We were nothing at all to anyone till you came among us; but at your house all are on the same level—we are not slaves."

There was a large congregation in the church the next day, councillors and others having come from headquarters. Mr Koyi conducted the service, and expounded the ten commandments, as we do at every church-service. I addressed the people, telling them of droughts in South Africa, and such as we have at home sometimes, and the services held by Christians every year to thank God for the harvest. I read Isa. lix. 1-8, and

connected that passage and Isa. lxviii. 6 with
Malachi iii. 10, from which I spoke, Mr Koyi
acting as interpreter. I pointed out what God
desired in place of sacrifices, and as they would
never think of praying to the spirits without first
sacrificing, so we had to learn from God's Word
how we are to prepare our hearts to seek Him.
A councillor who had killed a man just before
then was present, and as I read, "Your hands
are defiled with blood," he cried out, "He is
speaking out of his own head; that is not in
the Book." It showed, I think, that his con-
science was not dead. So clearly did the Bible
describe their thoughts and feelings that they
believed that we knew from it all their thoughts.

None of the warriors had come to the service,
and as they continued dancing at the royal kraal,
we determined to watch again at night. About
four in the morning slight rain began to fall, and
we retired to rest. Next day we had agreed to
hold another service to pray for rain, and at noon
the people collected, some of the chief's councillors being again present. At two o'clock, before
the meeting had dispersed, heavy rain fell. This
was the 18th of January, and seven weeks after
the rain in November. The incident made a
profound impression upon the minds of the
natives, and no doubt indirectly, if not directly,

advanced our work. The rain dispersed the assembled warriors, and the people again became engaged in planting operations, and quiet ensued for a time.

A few weeks after, on a Sunday, two councillors came to us with a sheep as a thank-offering for the rain. We refused the gift as we disclaimed having regulated the rain, and because, as we pointed out, they had sacrificed a score or more cattle to the spirits and received no rain from them, but confessed themselves beaten, while God, who had alone sent rain in answer to prayer, was to be paid by the gift of a sheep. They heard some plain-speaking and preaching and appeared glad when we allowed them to go, taking the sheep with them. The common people, who now began to be bolder in their attendance at the services, felt that we were being slighted too much by the councillors, and such an incident as the offering of the sheep was talked of far and near. It aided greatly in the furtherance of our interests, as all believed that by our prayers we could give or withhold rain, and considered that we should have accorded to us equal rights with the witch-doctors whose incantations had so signally failed.

The Sabbath meetings now became more firmly established, as the presence of the *amaduna* at the meetings held to pray for rain was taken by the common people as a recognition of them, and they were not afraid to come. The effect of the rain-question was to increase the interest of the people in the Book, and we were able greatly to extend our area of evangelistic work, and wherever we found the head of a village willing for a service to be held we visited his village regularly and preached. The attitude of the people towards us was more respectful and hearty, so we went on, rejoicing greatly. At the end of February there was a cessation of rain for about a week. Mombera had hanged a man for stealing cattle, and a deputation came to ask if we were offended at this and had stopped the rain. We again had the ear of the *amaduna* and tried to teach them the Word of God, and upbraided them for having left off attending the services.

That rainy season was a remarkable one, and the natives still remember and speak of it. Rain fell on one day in November, nine days in January, eleven in February, twenty in March, and four in April, *i.e.* on forty-five days, and only reached the exceptionally small amount of nineteen inches, yet the best harvest I have seen in Ngoniland followed. The rain fell in gentle showers and suited the character of the country. The natives say that they never had such a convenient rainy season, as it rained at night and did not prevent their work in the gardens during the day. The natives usually suffered from want of corn in the interval between sowing and reaping, as insufficient stores were made to carry them on to the harvest, and at the time of which we write, as the harvest was late there was great hunger. We had an opportunity of showing the Hoho people, who had been very troublesome and unkind to us, that we could warmly interest ourselves in their life and try to help them in time of need. We distributed a considerable quantity of beads among them, to enable them to trade with those who might have food for sale. I am afraid our kindness was not duly appreciated by all. The heads of the villages were called to the house, and beginning with the oldest we gave out the beads. The Chipatulas were consequently placed among the last and were very indignant and rude to us, as they considered we had slighted them in giving to others—and to slaves —before giving to them. The encouragement received from these men a little later on was not very marked. After their beads were used up, and the hunger still continuing, we offered to give them letters to our friends at Bandawe for which they would get loads of flour if they would send down their villagers, but we were told that we should get the Tonga who usually carry our goods to bring it up, and they would receive it. I do not quote these incidents as illustrative of all the natives, but for many a day it seemed that the people were unable to appreciate a kind act, and took it as an exhibition of our simplicity on which they desired to impose further.

In June we had to undergo one of our greatest trials when William Koyi was removed by death. Not till then had I fully felt the responsibilities of the work, or so great a sense of loneliness and helplessness among the Ngoni. In a brief biography of our friend I have tried to tell something of our loss by his death, and how I loved him, so that it is unnecessary to say more in this place.

While we were mourning the death of our comrade, Mr Williams and I were rejoicing that the restrictions on our work were being removed, and our position receiving more general recognition. It was while Mr Koyi was on his death-bed that there was a meeting of the chief, the sub-chiefs (his brothers), and their head-men. For some years there had been a feud between Mombera and his brother Mtwaro at Ekwendeni, and the permission we had asked to visit the latter had always been refused. As he was heir-apparent it seemed to us advisable to make his acquaintance, and we regretted Mombera's refusal. In the middle of 1886 the action of Mombera in having consulted us in regard to the rain, and seemingly having come under our power, stirred up the hatred of the other sections of the tribe. A desperate attempt was made by the disaffected in the council to overthrow the chieftainship of Mombera and openly follow their own ways. Again our prayers were heard, and after the turmoil of several days, the matter ended by Mombera and Mtwaro becoming reconciled, notwithstanding the opposition of some who desired the enmity to exist, in order to aid their effort to break up the tribe into sections. The four brothers pledged their friendship, and the kingdom was maintained intact. That and other matters were settled in open council, but the question of our presence and work was taken up in private by the chief, his brothers, and the councillors. This was no doubt owing to the presence of large armed escorts which had come with the sub-chiefs; in them the war instinct was active, and they were eager for the excitement of open discussion.

What was said and done in private we do not know, but we were informed next day by a deputation representing each party in the council, that we must understand that we were free to preach the Gospel, and teach the children in every part of the country. They expressed the hope that, instead of confining our work to the people around one station, we would open stations in each of the principal divisions of the tribe. We learned afterwards that Mombera was accused of receiving goods from us, and that the principal thought in their minds was, that by having a resident white man at each sub-chief's village they would also share in the spoil. There was full proof of this eighteen months afterwards, as will be seen further on ; but it was evident also that the growth of popular feeling in our favour was proving an uncomfortable fact in the mind of the chiefs, and they were compelled to open the country to us. I wrote home at the time as follows :—" I can point to no particular incident closely connected with the happy change in the feelings of the people ; but nothing more satisfactory can be said than that the cumulative force of the Christian life and teaching of those resident here has slowly but surely produced its natural effects on their minds. Various incidents, such as the rain question last January, could be cited as distinct stages of advance, but no part of our work has been without its power ; and I believe that the patient waiting of the past years will be amply justified and rewarded in the results of the future.

"They do not desire to engage in war, and the only advocate for war at the council meeting was shouted down by the assembled councillors. Since Mr Koyi's death a deputation of councillors came from the chief, on account of a rumour having been spread that since the country here seems to kill all our fellow-workers we would now leave. The chief sent them to say that we must not leave, but consider our position the same as if their special friend Mr Koyi had lived. To us a few days before Mombera said, 'I understand your work to be such that if deaths do occur it will still go on. God gave us life, and He can take us away when He pleases, and we cannot say aught.' We assured him that he had spoken rightly, and told him that though we should die there would be others who would carry on the work. Though teaching was proscribed, we have three youths reading the New Testament, and others coming on rapidly. Most of these are also earnestly striving to know God and walk in His ways, and from among these we will find helpers when we formally open school.

"Our position and prospects are cheering. Mr Williams has agreed to extend his engagement for two years meantime; but as he must now reside here the station at Chinyera occupied by him will be closed, except on Sundays, when one of us will walk over and conduct services at one or more villages. A good climate and extensive opportunities for service can be offered here, and I trust the Committee will be able speedily to fill the place of Mr Koyi. We should with another helper be able now to itinerate, which is a method of work which would be greatly appreciated here by the people. To fully equip the station, an ordained missionary should be sent, for we have hopes that a native church will be very soon established here."

Mtwaro had also a personal interest in becoming reconciled to Mombera and in professing an interest in us. Some time before he had sent a messenger to me requesting my presence at his kraal in order to treat an affection in his left knee-joint. I sent back the reply that I would gladly come to him if he would first obtain Mombera's consent, as I had been refused permission to visit him. He (Mtwaro) had heard of the medical work and desired the benefit of it in his own case. When we were permitted, as I have related, to visit Mtwaro I went to him, but medical treatment was unsatisfactory on account of the superstition of his head-men, who would not allow me to touch or examine the knee-joint. My visit enabled me to know the expectations of the people, and their begging for cloth was most irritating and trying, but was satisfactory in so far that the ice was broken and another door opened for our work. I was, however, not allowed into the village, but had to pitch my tent outside in the bush. In the middle of the night I found myself alone, with the hyenas sniffing round the tent at my elbow, as my men had crept away to the warmth of the huts. During the day the people crowded round the tent, and more than one hand could be seen pushed under the canvas at one time to pull out whatever they could grasp. With the exception of Mr John W. Moir who had visited him in 1879, no white man had met Mtwaro before at his kraal.

We were encouraged when harvest came round by finding among the people, in some of the villages where we conducted services, a desire to have a special meeting to thank God for the crop about to be reaped. They said God had given them the harvest, and they should thank Him for it before they began to reap. Thus for the first time in Ngoniland, on the people's initiative, a heathen custom,—the feast of first-fruits,—was replaced by a service of praise to Almighty God. It was the more encouraging as it came from the villagers among whom the Word had been longest preached, and was in marked contrast to the ignorant talk of those who were not instructed. A large and hearty service was held, and then they set about gathering in their crops.

In August I left for Mandala and returned with my wife in the beginning of October. The reception accorded us on our arrival was very warm, and an explanation was given of the scanty respect shown on some former occasions. The chief said, " Yesterday you were a boy ; to-day you are a man and can speak." The Ngoni accorded the privileges of manhood, such as transacting of business, to married men, and as long as I was unmarried it was contrary to their habit to have to treat with unmarried persons whom they considered to be boys. It is un-doubtedly the case that the married state has been more helpful to the progress of the work than the unmarried had been.

We were not long in starting a school when we obtained permission, and from the first we had two natives, who were able to read, to help us in the work. They were two of those who had been taught in the evenings and they proved a great help. After the first fortnight, the whole of the sixty children attending came and de-manded their pay for learning the Book. When they found they were not to be paid, they refused to come, and again the Chipatulas showed their hand in preventing them from coming because we also refused to pay them for allowing the children to come. The two native teachers, however, from among those in their village were able to collect twenty-two scholars, and so again the school went on, that being the number in attendance for nearly a year.

When the school was fairly started, Mombera sent the ominous warning, "You must not cultivate your garden merely in one place," meaning that the jealousy of the others would be aroused if we did not immediately begin schools in their districts. We explained that on account of distance that could not be accomplished until we were reinforced from home, and went quietly on with our work at Njuyu, making efforts to extend our influence in the new districts. Our efforts in the latter direction revealed how much the questions of war and cloth were mixed up with their talk about schools and preaching, and discounted their professed acceptance of our work. It was increasingly evident that we could not rely on political changes, or edicts of councils to establish the work among them, and we therefore bore with their ignorant conclusions and temporal expectations, and strove to have the spiritual power in the work which would establish and extend it. As we had been long in getting a commencement made in school work, we deter-

mined that the schools should be evangelistic
agencies, and the workers in them only those
whose lives were consistent with their profession.
The question arose regarding one who was a
polygamist, although in other respects consistent,
being allowed to teach, and his offer of service
was declined until he should dissolve his polygamous connections. In after years the wisdom
of this step was revealed.

The year 1886 closed with one school in progress and evangelistic work being carried on at
six or seven centres. A new era had begun in
Ngoniland.

CHAPTER IX

IN MEMORIAM : WILLIAM KOYI

FOR the following particulars of the early life of William Koyi I am indebted to *Lovedale: Past and Present*, and the account of a humble yet worthy convert from African heathenism will be read with interest. " William Koyi was born of heathen parents at Thomas River in the year 1846. His mother died a Christian. He left his home during the cattle-killing mania in 1857, and went to seek employment among the Dutch farmers in the Colony, earning half-a-crown a week as a waggon-leader. About this time his father died, and five years later his mother and two sisters. He left his Dutch employer and worked for five years at one of the wool-washing establishments at Uitenhage, and was promoted to be overseer. From thence he went to work in the stores of Messrs A. C. Stewart & Co., Port Elizabeth, where he remained for about the same number of years. He had never attended school, but now felt the need of education, and therefore set about learning to read Kafir. He had about this time, 1869, been converted, and been admitted a member of the Wesleyan Church at Port Elizabeth.

"He came to Lovedale in 1871, and his case is one of the most remarkable results of Lovedale work. A stray leaf of '*Isigidi mi Sama-Xosa*,' which he picked up and read during his dinner hour at Port Elizabeth, was the first cause of his attention being directed to the place. On enquiry he found it was 150 miles distant, and he then resolved to walk to it and seek admission. He had friends at Tshoxa, Rev. Mr Liefeldt's station, and it was from that missionary he brought a note of recommendation. He attended the first, second, and third year's classes; and during his stay at Lovedale he was active, willing and trustworthy, caring for duty and not for popularity among his fellows.

"He came to regard Lovedale as his home, and to be regarded as a humble but valuable worker who could always be depended on and needed no pushing to his work or pressure to keep at it and do his best, and make himself generally useful. After a time he was appointed assistant-overseer of the work-companies of the native boarders.

"In 1876 he offered, along with thirteen others, to go to Livingstonia as a native Evangelist; only four including himself were chosen. He has steadily continued these nine years, at the work at Lake Nyasa, and shown considerable energy and natural intelligence, and has thus proved to be of great service to the Free Church Mission in Central Africa.

"He returned to Lovedale for a time to recruit his health, and in 1882 married the second daughter of the late Rev. A. Van Rooyen of Blinkwater, Fort Beaufort: she also a little later proceeded to Lake Nyasa, and is now engaged there in the work of the Mission."

The foregoing account was in type in 1886, in which year William died on the 4th June; and soon after, his stricken widow, herself in bad health, returned to the Colony to her own friends.

William served the cause at Cape Maclear in its early stage, afterwards removing with the others to Bandawe. In 1877 he accompanied Dr Stewart on his exploratory journey along the west side of Lake Nyasa. In 1878 he accompanied Dr Laws and Mr James Stewart on their journey of further exploration of the west side of the Lake, and on their meeting with Ngoni on the hills to the north of Bandawe he was invaluable to the party, being able to speak their language so as to be understood. This was the Mission's

first contact with the Ngoni, and William was
the first to speak the name of God to them.   In
1879 he accompanied Mr John Moir in his visit
to the Ngoni and to the Basenga on the Loangwa,
several days' journey west of Ngoniland.   Later
on in that year he accompanied Mr Stewart on
his journey from Ngoniland northward to Karonga,
and westward to Lake Tanganyika.   If we remember that Stanley and other African travellers have noted how African travel proves a man's
character more than any other mode of life—and
they refer to Europeans—and think of those long
and arduous journeys of William Koyi, during
which his character stood the test, no more need
be said as to the genuineness of it.   Of him, Dr
Laws, under whom he laboured for several years,
wrote these words : " William has been a true-
hearted and earnest worker in our Mission ; and
in many a difficult time in dealing with the tribes
among whom Mr Stewart, William and myself,
were travelling, his advice and help proved most
useful.   In 1876 when Dr Stewart of Lovedale
was coming up to join us and be at a native
meeting, he called for volunteers to go with him
to Nyasa.   A number stood up, and last of all
William got to his feet, saying that though he
had not the education of the others, he had the
desire to engage in the Master's service, though

he could only go as a 'hewer of wood and a drawer of water.' Since then he wrote of having half a talent, but being anxious to use it for Christ. This spirit of humility, so alien to the tribe to which he belonged, has been honoured of God, and doubtless many will yet arise to call him blessed, having first heard from his lips the Word of Life."

It will illustrate the character of William Koyi if I give a few incidents connected with these. On one occasion, not long after the Mission had settled at Bandawe, report of a large Ngoni war-party, on its way to attack the people around the station, was brought from a village some miles distant. On such occasions the terror-stricken natives, women and children, rushed to the vicinity of the station in hope of protection by the Europeans. Thousands of helpless women and children crouched among the bushes around the station, or crawled into holes among the rocks on the neighbouring hill, or lay on the beach ready to take to the water as a last chance of life. On one occasion, not only were the natives alarmed, but so threatening were the circumstances that the missionaries hastily put together a few things and launched the boat ready for escape to the rocky island some hundreds of yards off. As Dr Laws was on the beach superintending operations, he was attracted by a little boy with book and slate in hand near to him. As nothing apparently could be done to save the natives, or the station, Dr Laws said to the boy, "Run away and save yourself," to which the little fellow, clinging to his only possessions worth saving, replied, "Where shall I run to, white man?"

When the report above referred to reached the station, a consultation was held, and Mr Koyi volunteered to go out and meet the war-party, and endeavour to turn it back from its purpose. He walked on for some hours, and at last met the party at a little stream, where it had made a temporary camp to await a favourable opportunity to attack the village of Matete, some two hours west of Bandawe station. It turned out to be a party belonging to the Chipatula family, before referred to as having been the first to receive kindly the Mission party in 1879. They were, it was stated, not only intending to attack the natives, but also the Mission station, in order to secure the wealth of cloth, beads, and other goods they imagined were in store there. When Mr Koyi met the party, and before he could open his mouth, the young warriors began to engage in war-dancing. On such occasions the slightest indiscretion in speech or movement, which might

Ngoniland Staff at Njuyu.

Hora Mountain—Scene of Tumbuka Massacre.

be interpreted as defiance, would have led to an immediate attack. There, with only a few friendly boys, William beheld the awe-inspiring war-dance of the Ngoni. They danced in companies and they danced singly, each warrior clad in hideous-looking garb which, with their large war-shields, almost hid their human form, and made them more like war-demons than men as they leaped and brandished their broad - bladed stabbing spears which they fight with. Mr Koyi stood for a time watching them, and utterly unable to decide what he should do, or how to effect the purpose for which he had come out. With secret prayer to God for guidance and success, he sat down on the bank of the stream. Still at a loss to know what to do, he took off one of his boots and stockings and began to wash his feet. That done, he, as leisurely and still puzzled, put on his boot again; but still the dancing went on, and there was no opportunity to speak even had he known what to say. He then proceeded to wash his other foot, and the warriors sat down. He found the opportunity for speech, and with his native instinct remarked, in an off-hand manner, " Now you are sensible people to rest yourselves on this hot day." This produced a burst of laughter from the warriors. The spell was broken ; the war-like intentions of the party were frustrated, and then free and open speech was found. The result was, war was averted and a section of the party was conducted to the Mission station, when it was arranged that Mr Koyi and Albert Nama-lambe, who was at that time at Bandawe, should go back with the party and see Mombera, with a view to a permanent residence among the Ngoni.

Thus, in the providence of God, the party that left home bent on war and plunder, returned home as guides and escort of the messengers of the Gospel of peace; and that incident, which well illustrates the valuable work of our departed colleague, was the prelude to the commencement of work among the Ngoni, the success of which has been phenomenal, as we shall presently see. Mombera once said to me, "My army, when away from home, are like mad dogs; they cannot be kept in, but bite small and great the same." And only those who passed through the fire of the pioneering days at Bandawe and in Ngoniland can measure the service done that day, not only to the thousands around Bandawe, but towards the success of the Livingstonia Mission. Years after, on encamping at that village near which the Ngoni army was met, the chief related the story to me, and sent with me for Mr Koyi a bunch of bananas to show that he had not forgotten what he had done for them.

When Mr Koyi accompanied the warriors back to Ngoniland, he and Albert were introduced to Mombera, and resided in a hut in his village. The Ngoni took some time ere a welcome was given; there was one party favourable to and another against their being allowed to stay. They were exposed to many insults and threats, and for a time their position was most critical. They could not both go to sleep together at night, but took turns in watching on account of the threatening attitude of the people. In all these times Mr Koyi's knowledge of the Kafir language was invaluable; and Mombera, despite his rough manners and despotic behaviour, was extremely fatherly and fond of children, and formed a remarkable attachment to Albert, who had a very attractive appearance and manner.

Mr Koyi was known by the native name of *Umtusani*, and from love to him Mombera named one of his sons thus, just as afterwards he named one after Dr Laws as *Robarti*. Mombera was very kind to Koyi, and although he only made sport of what was told him of the Gospel, he always showed him great respect, and became the butt of his head-men on account of his attachment to him. On the occasion of the last great tribal ceremony of putting crowns on the heads of those who were henceforth to take

their place as men in the tribe, there was a
gathering of several thousands of armed men
from the different sections of the tribe at the
royal kraal.   The crowning ceremony I elsewhere notice, but here I mention as showing
how prominent and open was the hostility to
the representatives of the Mission for many a
day, a clamour got up that Koyi should be
killed.   He was present in the cattle-fold, as it
was always found advisable to go about without
giving evidence of fear, as one of the best methods
of disarming their hostility.   One of the most
famous of the Ngoni generals, named Nawambi,
led off a great war-dance which Koyi described
as making his hair rise up.   This valiant's war-
cry was "Beka pansi" (submit).   His movements were terrific to witness, as I once beheld
them myself.   We were wont to call him Bel-
shazzar, for in his war-dance he "lifted up
himself against the Lord of Heaven."

With spear in hand he began by walking with
raised proud look round in front of his warriors.
Then kicking the dust of the ground over those
around, and pointing his spear in seeming indignation, said "submit."   The assembled thousands of warriors, beating their shields with their
warclubs, cried "submit."   Then he named the
surrounding tribes, the hills and mountains,

the sun, moon and stars, his seeming fury
waxing stronger and the clouds of dust flying,
while at each call the warriors beat their shields
and roared "submit." The elements of nature,
rain, thunder, lightning, were all called on to
submit; and amid the increasing din of shield-
beating and roaring of the warriors, the climax
of his dance and his daring blasphemy was
reached when, pointing to the sky, he cried,
as the foam flew from his mouth, "Wena
spezulu! Beka pansi!" no doubt meaning
Umkurumqango, the God they spoke of as
dwelling above. The tumult was as if all
assembled had turned into demons, and no
wonder great fear fell on Mr Koyi. Mombera
saw his discomfiture, and rising up, went and
took him by the hand, and led him to his own
place and sat down beside him. It was probably
what saved Koyi's life on that occasion, for once
a cry of blood goes out in a company of warriors,
fired by such dancing as that of Nawambi, they
indeed become as mad dogs or worse. Such
scenes have for ever passed away, but in those
days they always ended in bloodshed.

William was in perils oft. On the occasion of
a visit of Dr and Mrs Laws to Ngoniland, Mrs
Laws in a kindly manner put her hand on the
head of one of Mombera's children with the

remark, "Such a fine child." After they had gone the child sickened and died. The cry got abroad that he had been bewitched when the white lady put her hand on his head and re-marked on his appearance—a thing the people refrain from doing, reminding one of the super-stition at home connected with "for-speaking" anyone, especially a child. The matter was threatening enough at the time, and it reveals something of Mombera's character when he secretly informed Koyi, and said that he him-self did not agree with those who said the child had been bewitched. The matter was of great importance, and the council summoned the divining men who fortunately blamed some evil spirit and not Mrs Laws. The council was not satisfied, and more than likely the party op-posed to the Mission conceived the idea of seizing on this as a pretext for driving Koyi out of the country, if not of killing him. Secretly Mombera informed him of all that was going on. The coun-cil insisted on having recourse to the Tonga *muave* ordeal, and so fowls representing the Mission party had the poison administered to them. They all vomited, which had to be taken as evidence of the innocence of the accused. But so deter-mined apparently were the council to obtain a conviction, that they suddenly discovered that the usual test as to whether the doctor presiding was giving true muave or not had not been carried out. Another fowl was therefore taken and received the poison and died. This shows how insecure for a long time was the position of William Koyi and the others.

These were not the only occasions on which our colleague was placed in trying circumstances which required great wisdom, manliness and devotion to duty, but all through there was no wavering or weakness shown. He understood his position and the trust which was placed in him, and with characteristic humility and absence of self-seeking, he went through it all, counting it an honour to be a messenger of the Cross to the Ngoni. A European member of the Mission once said to me, " It requires great grace to be humble, when one is called *Mfumu* (chief) by the people on every hand." If a European with his education and attainments found himself tempted to be lifted up by the merely respectful greeting of the natives, how much more so might Mr Koyi be expected to feel that temptation, in the position assigned to him in Ngoniland and the respect and affection of chief and people which he gained for himself! Those who have had to deal with natives understand how many a native, otherwise good and trustworthy, loses himself entirely when intrusted with a little authority.   But Koyi never forgot "the hole of the pit whence he was dug." The character for steadiness, humility, and devotion to duty, which Dr Stewart gave him, was fully borne out to the very end.   In those early days Mr Koyi had to bear the chief burden of those frequent outbursts of Ngoni pride and impatience.   If he was not there alone and having to meet them by himself, he was, till near his death, required as interpreter and chief speaker.

I became aware on several occasions that he hid from others and from me much of the anger, hard words and wicked intentions of the Ngoni. He was, as a native, able to discount what they said, but the kindly nature of the man was shown in his rather suffering obloquy himself than that his white friends should be distressed. This was shown on another occasion. During a time of trouble, when we were being accused of inciting the Tumbuka to revolt, there was great distrust of us manifested. It was a Sabbath morning, and earlier than usual some people were gathered for the service. Some head-men and others fully armed came over from the chief's villages, as they said, to pray to God. This was very unusual, and as we knew it was reported that the attendance of the Tumbuka, who were coming on Sabbath to our service in large numbers, was exciting the jealousy of the Ngoni, the presence of armed men led Mr Koyi to apprehend trouble that day. To add to his view of the situation, from the hollow below the station, between it and the chief's residence, we had all morning seen smoke arising from a number of fires. Mr Koyi asked the armed men who came from that direction what it was, and they said some people were roasting cassava there. After observing Mr Koyi's restlessness and troubled face, I asked what was causing it. He then told me that he feared trouble at the service, and proposed that I should remain in the house and not go to the service that day. I said that could not be, and we went to the service together, and Mr Koyi preached. Everything passed quietly except that in the middle of the address a leading man got up from his place and gathering up his spears said, "We have heard enough of that. Give us cloth. That is what we want," and walked out alone. The others seemed ashamed at his conduct. At the close of the service William came into my room, and with a half-ashamed look on his face said, "Did I not give my knee a great knock to-day?" This was his parabolic way of saying that he had been frightened at his own creation. He explained it by relating how a Kafir, tired while on a journey, had lain down to rest and fallen asleep with one of his knees flexed. On half awaking he saw the knee as if another were over him ready to slay him. Reaching out for his knobkerrie he dealt a blow on the supposed murderer, only to find it was himself he had hurt. This, I think, was the only occasion on which Mr Koyi showed that his fears were near unmanning him, and to Africans the matter is plain when I say he had been suffering for some time from feverish attacks. It appeared, however, as we afterwards learned, that the head-men had indeed come to hear what was said at our services.

Although little has been said of it above, Mr Koyi was a devoted evangelist, and so far as liberty to carry on such work was given, he was eager to embrace every opportunity of telling of the love of Christ. He preached by his life, and to a great extent, and with an effect we shall never know, his personal talks with the people were powerful means of keeping our real work before them. He was a diligent student of the Word of God, and with much of the warmth of Christian feeling, he was a happy Christian. He had persevered so as to acquire a fair use of the English tongue and literature. A common Kafir—a Mission Kafir—

to be sneered at by men not possessing a tithe
of his manliness or good character, he was one
with whom it was a privilege to associate.   I
acknowledge with pleasure, I received un-
measured help from him; to his achievements
in those early days the after-success of the
work was in a large measure due. He died
before he saw the fruit of his labours among
the Ngoni.   He lived in the assurance that
the day would come soon when the work would
be allowed to go on unhindered by the council,
and he had a large idea of the importance of
gaining the Ngoni, so that in his letters to
Lovedale he showed himself as he was on that
subject.

He could take a comprehensive view of the
aims and work of the Mission — looking beyond the immediate future to a degree which
was most remarkable for a native, and which
exceeded that of some of his white brethren.
He strongly urged upon his fellow-countrymen
in the colony the importance and character of
the work, and the call to them to give themselves to it.   Writing home in 1883 he says, "It
will be a great day when the native Christians
of South Africa will willingly undertake the
work here, and give up their lives to come and
teach their countrymen at Lake Nyasa.   I wish

I had a better education; I would give myself wholly to my countrymen here. Here is work for Christ standing still. You (native Christians) have received much, and have also received education. I do not say you do not work with that education where you are. But can you not even spare two to come and teach these people who are dying in darkness? What am I to think, and what encouragement will my soul receive if no attempts are made by you to second my poor efforts? My great wish is that there was a white and also a native missionary here, and then the work would progress. I think there should be more coming to help in this great work." That "great wish" was the conviction of Dr Laws also, and my being sent out in 1884 was the response to it by friends in Scotland.

And his death? How died the faithful soldier of the Cross? As he had lived, strong in faith and in the assurance of acceptance with God through the merits of Jesus Christ. The sickness of which he died ran a rapid course. Having to go to Bandawe, I left him convalescent from an attack of malarial fever. I had been away only a few days when his condition became serious, and he expressed a desire to have me with him, so I hastened back to find to my dismay that a dangerous affection of the heart had supervened. He rallied for a time, and though still confined to bed, he was full of hope that he was to be raised up again for his work. One day towards the end a large deputation came from the chief. As they were seen ascending to the station we were anxious as to what its object might be, having only too good reason from past experience to be anxious. Great was Mr Koyi's regret that he could not take his wonted place when the deputation arrived. It was the happiest day of my life—they had come to say that we had now full permission to teach the children and to go about the country. No sooner had the deputation withdrawn than I hastened to the sick chamber to give the good news. As I entered, William, who was sitting propped up in bed on account of his laboured breathing, said eagerly, "What is it? Can you believe it?" I said, "We have now full liberty to carry on all our work, and to open schools." Clasping his hands and taking up the words of the aged Simeon as he beheld the Saviour, with a never-to-be-forgotten gleam of joy lighting up his wasted countenance, he said, "Lord, now lettest Thou Thy servant depart in peace, for mine eyes have seen Thy salvation." He was overcome, and lay for a time as if dead.

The words he uttered were his prayer, and

it was answered two days later, when in peace,
and with a brief farewell to his wife and myself,
he was taken to the higher service of the sanctuary above.   The words he uttered were also his
thanksgiving and his resignation.   During the
interval till his death, quite contrary to his
former hopefulness of recovery, he was assured
he was to die.   He once said he would like
to be raised up to see the work in progress,
but he knew it was to be otherwise, and he
said it was best.   So died William Koyi,
having been a humble and faithful follower of
the Saviour, a trophy from heathenism, and the
pioneer of the Gospel in Ngoniland.   It was
meet that, his work done, his dust should rest
where he had fought the battle, becoming the
title-deed to "Ngoniland for Christ."   His was
the second mission grave opened there.

Mr SUTHERLAND, Artisan Missionary.

CHAPTER X

IN MEMORIAM : JAMES SUTHERLAND

ALTHOUGH less closely connected with the work in Ngoniland than William Koyi, it is fitting that the name and work of James Sutherland should be had in remembrance also, for he was the first European missionary to reside among the Ngoni, having been sent to be with Mr Koyi in 1882, and along with him had to bear the trials of those early days.

I write not merely as a fellow-worker but as a close friend, and having had much to do with his choice of foreign mission service. Born in Wick in 1856, he was converted, when a youth of eighteen, about the time of Moody's visit to Scotland in 1874, and became a Sabbath School teacher, and "Monthly Visitor" tract distributor in his native town.

In 1876 I went to Wick as missionary to the fisher population. I gathered around me a band of young men, young in years like myself, and young in the Christian life. They became my best helpers, and along with me took part in open-air and in-door meetings.   We formed a class for the study of the Bible, Christian Evidences, Butler's " Analogy," and such-like subjects, as well as for reading in the English classics.   The friendship then formed lasted to the end.   As I then looked forward to foreign mission work myself, the subject of missions was well discussed, and it turned out that I, who had fostered the desire in his mind, was beaten in the race to Livingstonia by four years.

Following his father's trade—that of a shoe-maker—he was a great reader and worked steadily to improve his education with mind bent on higher spheres.   He was particularly drawn to scientific studies, and gaining a bursary he entered Edinburgh University along with one of his companions (a member of my class), who, like him, had been fighting his own way in the world.   It is characteristic of their thirst for knowledge under difficulties, that, both being shorthand writers, they each took a different class outside the line of their special studies, and by interchanging their note-books had practically the benefit of both classes for one fee each.   After two Sessions at Edinburgh, in 1880 a man who understood agriculture was wanted for Livingstonia.   The advertisement was eagerly read by

the two students, and as agriculture was one of
the subjects they had been studying, each determined, unknown to the other, to apply. A late
member of the Mission, studying medicine along
with me, was deputed to see the candidates, and
by the same mail I received letters from them
both asking for a recommendation. James
Sutherland was chosen, and the other went to
India to a commercial life, but was no less a
missionary. Both died within a few months of
each other.

Leaving home in the middle of 1880 he was
for some time engaged in the work of the Mission
at Bandawe under Dr Laws. He had been
appointed to fill the post of agriculturist, rendered
vacant by the death of his predecessor when on
the point of leaving for home at the end of his
engagement. But while nominally agriculturist,
he was, like all the others on the staff, everything
by turns, as was rendered necessary by the conditions of work in the beginning which was being
made at Bandawe. He was engaged in the erection of manse and school; in the laying out of
the station and garden; in testing the capabilities
of the soil at and around Bandawe for the
development of agricultural work; and at times
had to take his turn at school and meeting.

He saw Bandawe founded, and it was there he

was permitted to do most evangelistic work for which he was eminently suited. His close contact with the people every day while in charge of out-door labour enabled him speedily to master the Tonga language. He gained the affection of the poor down-trodden Tonga people by his happy disposition, and sowed much seed of the Word, which has no doubt helped towards the success of the work there since. Whatever his hand found to do was done with all his might. His sympathies were broad, and with the various phases of mission life and experience he was in harmony, keeping ever before him the great end for which he was sent out. He felt the troubles and anxieties consequent on the Ngoni raids long ere he was sent to be among that people. On one occasion he was lying helpless in fever when an Ngoni army was reported to be marching on the station. The sudden change from the quiet of the neighbourhood to the tumult and cries of the frightened natives, and the hasty preparations for flight of the Mission party, was a great strain on Sutherland in his weak state, and in a single night his hair began to turn grey. But he never regretted having given himself to the work, and it would be difficult to picture the harmony, happiness, and at times the mirth of the three youths who lived in " Bachelors' Hall " as they named their house. In those days the conditions of health were not very good, and long periods passed without a mail or news from the outer world, but he and the others lived and laboured as if all depended on their exertions.

When, in 1882, he was sent to Ngoniland as co-worker with Mr Koyi, he set himself to learn the language of the people, but by means of Tonga he was at once at work among those slaves of the Ngoni who spoke that language. There, as at Bandawe, his influence was chiefly among the common people, and many in bondage to the Ngoni had a new feeling aroused by his kindly words and the telling of the story of Christ's love. He spent much of his time in the neighbouring villages, and gathered under his influence a number of young men, many of whom have since become members of the Church, and some are now respectable members of society, who were, before he took them up, wild reckless youths bent on following the ways of the Ngoni. At Njuyu he erected the brick dwelling-house. It was no light matter to begin such an undertaking, where, with the exception of brick-mould and trowel, everything else requisite had to be got by the labour of natives. The natives had probably never before seen a brick, much less moulded one. They knew to set a few sticks in the ground in a circle and tie over them a few wattles and some grass, which served them as a house. Besides, the people had no inclination for work, they lived, not by the sweat of their brow, but by stealing the things of others. With such a set of helpers Mr Sutherland had to start to build a house of several rooms, which, according to the frequently expressed view of the natives, when finished, was a village under one roof. The clay had to be dug and puddled, then, by means of moulds made of disused provision cases, shaped into bricks and laid down in the shade to dry. For a long time it was extremely irritating work to spend a whole forenoon in teaching two or three to mould bricks, or others to lay them down flat in rows in the shade, and find when the bell rang for mid-day rest the whole squad of workers demand their pay, saying they had now worked a great deal and needed to rest for a time. How could a hundred and twenty thousand bricks be made at that rate? But Sutherland struggled on, and erected a noble house. Although only of sun-dried bricks, which are easily dug into by white ants, it stands to-day a monument of taste and thoroughly honest work.

It is a common experience in many Missions, that some men with special handicrafts engaged to do special work in the mission, develop the idea that mission-work is only preaching and teaching; they despise their position and mis-judge their influence, and in time throw up the work, or remain and are a source of trouble to their colleagues. Sutherland was not one of those. He made all his work true missionary work by his consecration to the service of Christ. Often did he turn up the pages of his shorthand note-book and read over a sermon preached by Dr Laws on Zech. xiv. 20, "Holiness to the Lord," treating of the ideal and possible in even the manual labour which might be engaged in for the Mission. " Let the spirit of every one impress even on the bricks he makes the motto, " Holiness unto the Lord !" Such was Sutherland's aim in all his work, and was that which enabled him to live a tranquil life amid many worries. He could have answered the speaker, who, at a meeting of the Anthropolo-gical Institute reproached the missionaries as do-nothings, and called the natives *imitative brutes,* in the words of Zimmerman, " Did I not put tools into the hands of the natives, and teach them to fell timber, to saw boards, and to make them into doors and window-frames ? Did not I myself dig the clay and make the first hundred bricks, in order that the *imitative brutes* might do the same? Did not I dig the ground and build the foundation walls of brick and mortar, until I could trust *these brutes* to proceed by themselves? Yes! I have now a house which shelters me, and compared with the sheds of the natives, is more like a palace. You say the African is like the ape, an animal gifted with the power of imitation. Well! Only his power of imitation goes a little beyond that of the brute."

Ten years have made a great change on the working habits of the people, and it is even yet not an easy matter or one conducing to equanimity of temper, to have to superintend natives at work, and what it must have cost those who laid the foundations of the present progress may be imagined. Sutherland had a sweet disposition, and was the right man in the right place. No one who knew him will forget his quiet behaviour under great provocation, and he had the happy knack of stimulating the native to honest work, not by fear, but by a powerful personal influence.

He was much respected by Mombera, and his happy manner made him attractive to many of the Ngoni. His life at Njuyu was too circumscribed for his ardent spirits. When building work was over, and as no schools were permitted, while it was unsafe to move to a distance from the station, he felt the hardship of not having his hands full of work. Added to this there were the frequent rumours of war, and excitement over the ill-concealed bad intentions of the Ngoni towards the Mission, and what was a daily annoyance almost past endurance — the begging of the people, from the chief and his wife downwards. It was not the polite request by one in need, but the insolent demand, and a volley of abuse if the request was not granted. One could scarcely name a thing that was not coveted and demanded. From early morning all through the day till near sunset there were people begging. There was no privacy, for they forced themselves into the house, there being only reed doors at first. On one occasion, in order to let a sick member have privacy and quiet, his bed had to be set behind the door to block the entrance. Mombera was often impudent and unbecoming in his behaviour, and so all the people were encouraged in the same manner of treating the mission-party when their demands were not satisfied. There was one good quality in the Ngoni character. If the missionaries were absent there was no attempt to enter the house. Their own laws were severe on that point, and death was the penalty for house-entering—it could not be called house-breaking, where the only lock to a native door is a cross stick outside, to which the door is fastened by a string.

Acting on the knowledge of this, when Mombera and his retinue, or a bevy of his wives—an impudent, drunken set of beings—were seen ascending from the river to the station, Sutherland and Koyi would hastily pocket some food, and putting up the reed doors, slip out at the back and scramble up Njuyu hill, behind which, in peace with a good book, they looked down on the begging visitors finding their journey in vain. Time was nothing to them, and they could as comfortably smoke, snuff, and talk gossip on the mission-house verandah as at home, and sometimes they remained a whole forenoon. It was an experience which could not have been avoided. The people were eager to get cloth and beads, but as for desiring the Gospel there was no evidence they did so, but much to show they were utterly opposed to it. In such circumstances it required tact and consecration to do as much Christian work as was done. While the prospect was of the most unpromising nature, Sutherland never lost hope or faith, and it was a solace in his many worries and unsuccessful efforts to reach the people, to gather the few house-boys round him every day and teach them to read the Word of God for themselves. In the early stage of such a mission, what people at home imagine is mission work may have but a small place. The foundation builders have as arduous a task, and of them is required as great faith and earnest work, though all they do will be hid by the superstructure raised by those who come after them, as those who, like myself and others, have been able, on what they achieved, to go forth to sow and to reap—to do such work as some are pleased to count as alone mission work.

The circumstances under which Koyi and Sutherland laboured were such that very little opportunity was given for, and little dependence could be placed on, oral teaching. Many a young, vigorous missionary, fresh from home, full of his own perceptions of the truth and his new duty, goes from village to village " bearing witness," and returns home feeling he has fulfilled his mission. But real missionary work in the early stage of a mission to such a grossly sensual, barbarously cruel people as the Ngoni, or their even more degraded slaves, the Tumbuka, more frequently consists in a consistent, loving life, than in sermons or addresses however eloquent. Such work is harder than preaching, and such work was well done by the departed brethren.

Like his fellow-worker, Mr Koyi, he was called
to the higher service above ere the fruit of his
toils in Ngoniland was gathered, but to-day, as I
scan the faces of those who sit at the Table of
the Lord in the church at Njuyu, I see one and
another who recall Sutherland to my mind, and
I can trace their spiritual history and meet him
in his dealings with them far back in 1885. The
worker may fall but the work goes on. So determined was he to cling to Ngoniland and live a
missionary's life, if he were not permitted to do a
missionary's work, that in 1885 when trouble for
us was abroad, and it seemed as if we would
indeed be driven from the country, Mr Sutherland was prepared to become a slave to them in
order to be allowed to remain. He even went
and chose his owner—an old and much respected
Swazi woman, the widow of Chipatula.

He packed up his goods, and on August 10th
left for Bandawe. On September 29th, 1885, he
died of haematuric fever within a week or two of
the expiry of his first five years' service. When
the news of his death reached Njuyu, the natives
came in large numbers to express their sorrow,
for *Sutherlandi*, as they called him, had won
their affection. Mombera also sent messengers
to speak his sorrow. As this was the first death
of a mission member he had known, he sent also

to ask if we believed people died by witchcraft, and if we thought our friend had been killed by the Tonga at Bandawe, he would set the matter right for us by sending down a war-party. It gave us an opportunity of speaking plainly to Mombera on the lesson for them of the life of Sutherland, and thus in his death as well as in his life he preached to the people.

CHAPTER XI

THE CRISIS: WAR, OR THE GOSPEL

THE year 1887 opened bright for us. We had a little school of twenty-two scholars and a class of young men too old to attend school, but who were anxious to learn the Word of God. It was now that the results of the past years of contact with the people could be estimated, when any who wished might come to us for instruction. For six years there had been nothing seen of what critics of mission work would term results, and yet we were now gladdened by observing that behind the apparent indifference of the people, and their merely worldly interests in clinging to us when we had work to offer them, the influence of the daily life of the staff had produced a marked effect. Although evangelistic work had been forbidden, the hundreds of workers who were engaged with us in brick-making, house-building, and road-making, formed an audience to which we ministered. Now when they were at liberty to come to our classes they did so, and, apparently quite suddenly, there arose a band of young men who were ready to stand by us. There was at least a mind open to what we taught, and their belief in some of their own customs was considerably shaken.

The youths who began to speak of our work as worthy the attention of the people, excited a violent storm of persecution. It was made plain to us that the edict of the chiefs, however favourable to our work, could not, and did not, change the natural mind of the population. It began to be a fight between the Gospel and the word of the witch-doctor, and the enmity of the natural and unrenewed heart of man. But to us it was full of encouragement, as it showed that these youths were not acting one thing in our presence, and another thing when among their countrymen. The leading boys in the prayer-meeting were sons of a witch-doctor who lived near the station. He also had come under the influence of Mr Koyi, and we were gratified by a step which he took in the beginning of the year. His village was the nearest to the station, and almost every day people came to consult him. The still mornings resounded with the responses of the applicants as they followed the doctor in "smelling-out" the case, and at times with the sound of his drums and the accompanying plaintive songs as some demon was being exorcised. When he began to perceive the nature of our work, and witnessed the effect of it on his boys, he moved his " consulting-room " to one of his other villages, about seven miles away among the neighbouring hills, in order, either to be away from the light which revealed his darkness, or out of respect to us. I believe it was from the latter, because about this time he discontinued taking with him, for the performance of his incantations in the country, his children who were attending our school and classes. In connection with these functions he required the beating of a series of drums of different pitch, and his sons and daughters were accustomed to do the work, but now he chose others and set them free. Wherever this man went he had a good word to say on our behalf, and his faith in us was further shown by his sending his wives and children to the station to be treated when sick. Yet although he attended our services and encouraged his children to cling to us, he continued to practice his profession, and his case shows how difficult it is for a native advanced in years to give up his long-established beliefs and follow a new course.

In marked contrast to that, and showing how important work among the young is, the case of

one of his sons may be stated.   He was supposed
to have become possessed of an evil spirit
(*chirombo*), and his father arranged for a dance
to exorcise it.   The son gave a passive obedience
to the arrangements made by the father.   I was
staying at a village a few miles from where the
dance was to take place.   On a Sunday morning
a woman, who is now an active Christian worker,
came to me to ask some blue cloth in which to
clothe the subject of the dance, it being supposed
that this *chirombo* was in quest of that variety
of calico.   I of course refused, and a few hours
later I received a note from the young man
Chitezi, as his name was, requesting me to send
him my Zulu Bible, as, while he had to submit
to his father, he desired to show that he did not
believe in what was going on.   In the evening I
went to see him and found his father, painted
with red clay, in the midst of his divining instruments, and in a circle around him and his son,
who sat reading the Bible, the drummers and
dancers performed.   It was a strange sight.
Such dances and performances were common
enough in the country, but never before where
the subject of them sat reading the Word of God.
Parental anxiety was no doubt shown, and on the
son's part filial obedience.   The one was not able
to exercise implicit faith in God and the Gospel,

and hence fell back on that which gave his mind
rest ; and the other was not strong enough to
declare a separation from the superstitions in
which he had been brought up.

Not long after this, however, Chitezi did un-
mistakably confess his faith before men, in deny-
ing and opposing the pretensions of an old
woman, who, as a " chief of hades " (*fumu wa
pansi*) practised her deceptions upon the community.   Such individuals travelled through the
country, dressed grotesquely and painted with
white clay.   They were credited with the power of
turning themselves into ravenous beasts, such as
lions and leopards, and of devouring any who
might incur their anger, or whom they might be
hired by anyone to destroy.   It was also believed
that the spirit of some dead chief was located in
them.   In consequence of this reputation, when
they turned up in a village in the evening, the
people were so frightened that they endeavoured
by gifts of cloth, beads and food to gain their
good-will and so be left unharmed.   Chitezi, on
one occasion, turned on one of these deceivers and
challenged her to turn herself into a lion.   Instantly the whole community turned on him and
affirmed that he was mad, and that his action
would enrage the " chief of hades " and bring
trouble upon them all.   He challenged her to

appear next evening at sun-set as a lion and he would fight her. She accepted the challenge. Chitezi, who, during the day was not quite at ease, armed himself with spears and sat on the village ant-hill to await the issue. Of course no lion came, but among the villagers the force of the incident was minimised by the woman's having gone away, leaving a message that out of respect to his father she refrained from hurting a child of Kalengo's. At the same time the matter was talked over and good was done by it. Only enlightenment of mind can remove the terrible fear which possesses them of what may happen if certain things are not done. It is not easy for them to give up their faith in their own practices. They are part of their life, and hence we find many instances where a patient being treated by us is at the same time undergoing their own treatment. In other cases, which at first almost comprised the whole of our medical work, it is only when their own doctors fail that we are called in. A medical missionary's most important work, or the ultimate end of it, is not merely to cure the patient. What his purely medical work greatly aids in accomplishing is the correction of error, physical, mental, and moral, and so he is compensated for the frequently unsatisfactory medical results of his efforts in many cases. The people cannot be laughed out of their (to us) absurd positions and beliefs. Their emancipation is a progressive work towards which all our work tends. In the Highlands of Scotland, and in the lowlands as well, I have met with instances of a blind obedience to superstitious usages, as firm and as absurd as may be met with in Central Africa.

When on this subject I may relate an incident which led to my having great freedom from molestation by those possessed of evil spirits. Some individuals went, or were led, about the country by their friends, supposed to be "possessed." At one time they affirmed the spirit could be exorcised by red cloth, and at other times by beads. They were usually well laden with such, but still the *chirombo* kept possession, and many overtures were made to me to help the cure by gifts. One day a strong, one-eyed man, named Luguta, whom I well knew as a bad character, was brought to the station. He fell down, and writhed and roared until the perspiration flowed from every pore, and the foam fell from his mouth. It was certainly a hideous sight, and well calculated to move anyone to pity if it had been free from deception. There was a band of young women with him, clapping their hands, which, they said, helped to quiet the spirit. They pleaded with me to give him something and let him go. I spoke to Luguta, but got no reply. My offer to give him medicine was rejected by the girls at first. They said the spirit wanted beads; but I obtained their consent, and went inside for my strong ammonia, which I applied to Luguta's nostrils. It put an end to his deception, and he ran off, not desiring a second inhalation. I said to the girls, "If you hear of any more evil spirits bring them to me, as I have medicine which they cannot stand."

Four years before this a wattle-and-daub school had been built in the hope and belief that ere long teaching would be permitted. This large, empty house was often the occasion of ridicule by the chief and headmen, who proposed that we should keep our cattle in it. When they were told that we had built it as the school in which their children were to be taught, they asked how long we would wait — till their beards had grown grey? We were told that the white ants would destroy it many long years before it should be required. We said quietly, "In that house your children shall be taught while they are yet children." These words were repeated on many occasions, when there seemed no likelihood of their being realised; but yet it came to pass, although, as we had said we would do, we had to keep it propped with trees inside and outside. We had triumphed and were glad, yet it was with a feeling of regret that one morning we saw the house collapse, and we were left in the middle of the rainy season with no house in which to conduct school. The verandah of the dwelling-house, however, served the purpose, and for nearly a year we met there.

We set to work to build a brick school. I found, however, that when I was engaged in school-work the workers in the brick-field did nothing at all. Instead of turning out twelve hundred bricks a day, we got only four hundred. I determined to give out the work in contract to one of the best men who could read, at so much per thousand bricks. He was overseer under me before, but was as dishonest at work as the others. I pointed out to him how he could work to his own advantage, and that if several workers could be got to work steadily, all the others would follow. He put his wives among the workers at the different jobs, and by what means we need not inquire he got them to lead the others. The work went on. He had more than double his former pay, and we had bricks at two-thirds the former cost. This was, I think, the first native contract for work in Livingstonia, and I was freed from attendance at the brick-field, and could devote myself to other work. When the brick-making was resumed after the events to be narrated in this chapter, a contract was not given. I pointed out to Chitezi that he and the others had now given proof of what they could do when working faithfully, and that as he was learning the Gospel he would understand what he should do in all his work. I showed how his former work, which yielded us bricks at a cheaper rate than before, was still too dear for the Mission, and that he would merely be paid as overseer; but if at the end of the season he worked satisfactorily, and relieved me as before, he would receive a bonus. This arrangement he understood to be equal to the ordinary standard of payment all over among the whites. Although it was considerably less than he had received under contract, yet to his credit be it told, his work was satisfactory, and he received his bonus. He has continued ever since as a worker in connection with the Mission, although polygamy and occasional instances of drunkenness barred his way into the Church.

For the first half of the year the work in all departments went on smoothly, and we had an ever-widening circle of adherents. The children made fair progress in school, and the station and village services were attended by people from near and far. Mrs Elmslie had got a fair start made among the girls by taking some into the house for training, and by class-work in the school, and a special sewing-class to which about twenty children came.

The only thing which arose to mar our happiness in the great change resulting from getting liberty to carry on the work, was the murder of six of our Tonga carriers on the road to Bandawe, by a band of Ngoni under Nawambi, one of the most notoriously cruel and indomitable warriors in the country. The late Mr M'Intyre, teacher at Bandawe, had come up to recruit after illness, and the Tonga who carried him and his loads to Njuyu were on their way home when they were set upon by the Ngoni in the forest. We were accustomed in those days to long intervals in our communications with our friends outside, as we were dependent mainly on Tonga to act as bearers of letters, so it was about six weeks after the event before we knew there was truth in the rumour that we heard in Ngoniland. We tried to get Ngoni to go with letters to ascertain the cause of the long silence of our friends, and our fears were increased when all refused to go down to Bandawe. At last two slaves, who had formerly lived at the Lake, and to whom as we shall see we afterwards owed very much, agreed to go. On their return we found that six innocent, industrious Tonga had been killed, and our carriers saw their skeletons lying near the path. When we tried to get Mombera the chief to take up the case he declined, not being anxious to try such a warrior as Nawambi for the offence. We even failed to get from him a condemnation of the attack, or any pronouncement which would tend to secure for us and our carriers a reasonable measure of protection. Nothing was said against us or our work, and we tried to live down the clamouring for war which the incident had markedly stimulated. We had our school and other work going on as before, but our faith in Mombera's former protestations of friendship was considerably shaken, and we observed that the attitude of the leaders of regiments, and many others, was less friendly to us than before.

We were well inured to trouble and anxiety, but the continuance of anxious days and the approach of a gathering storm told upon both my wife and myself, and we had a succession of attacks of fever which no treatment seemed to abate. Removal to another district in Ngoniland, although to live amid the discomforts of a native hut, quickly restored us both. This was the first occasion on which I proved the truth of Livingstone's advice, to move a patient to a different part when in a low state from which nothing seems able to rally him. Often, afterwards, good resulted from a change even from one house to another, or from one room to another in the same house. The natives have their own explanation of such a thing. They often carry a moribund patient to another village, or out into the bush, with the idea of cheating the evil spirit attacking him. The patient recovers, and they consider that they evaded the spirit. They always have a reason for everything they do, and in this connection I might mention that my wife when almost gone on one occasion, being exhausted through a severe illness, was saved by being fed with raw beef juice. The natives knew that she was apparently dying, and were tenderly sympathetic. When I had a bullock killed they knew it was for her, and the rumour went round that despite all my preaching I did exactly as they did, and sacrificed a beast to our ancestral spirit, and my wife recovered. Such things gave opportunities of meeting their difficulties and of leading them out into the light. Medical work was slow and unsatisfactory for a long time, just because it professed to be natural and not supernatural. Had we pretended to superhuman wisdom we should have had a much larger following in less time.

The explanation of the attack on the Tonga carriers and the altered disposition of Mombera and the Ngoni, before referred to, came out in the middle of the year in connection with a visit which Dr Laws paid to us in July. This was the first occasion after his return from furlough on which he had visited Ngoniland since 1883. He was put by the Ngoni *in loco parentis* to the whole Mission, and hence arose many of our difficulties, although to that, I doubt not, we also owed some degree of safety. At the time of his last visit to Mombera Dr Laws was requested to bring back with him bulls to improve the stock, woolly sheep, cloth, beads, brass wire, and even dogs, without giving him means to do so. During the years that elapsed till his return, his expected visit was often the subject of conversation among the people, but it was always in connection with the wealth that he was expected to bring to them.

When Dr Laws visited Ngoniland in July he gave a handsome present to Mombera and to the Chipatula family, yet he was received with but scant courtesy by Mombera and his headmen. It was evident that they were dissatisfied, and when reference was made to the slaughter of our carriers a few months before, Mombera and others made defiant charges against the Tonga, and no satisfaction could be obtained. Mombera would not listen to any serious talk, and all reference to our work was ridiculed. Dr Laws remained a week with us, but Mombera, contrary to his custom, paid no return visit, but sent begging messengers daily to the station. It was evident to us that there was a storm brewing, but we could not understand it. About a fortnight after Dr Laws's visit rumours of discontent were afloat, to which we paid no attention, until Mombera himself spoke to us on the subject. From various conversations with the chief, we became aware that the agitation had not arisen in the part of the tribe among which we had lived and laboured, but in the districts of the brothers of Mombera, which had never been overtaken, and over which our presence and work had consequently exercised no influence. We had no doubt of this, and the reason for Mombera's action was that he was being harassed by his brothers, and blamed for keeping to himself all the missionaries and the wealth that they imagined we bestowed. We were told by Mombera that Dr Laws would require to come back and settle the questions which were agitating the minds of the Ngoni. These were : 1. Their Tonga wives and children who ran away some years before had not returned as was expected when the Mission began work ; and as there was no expectation of their doing so, they had decided that if we could not bring them back without war, they would fetch them from Bandawe for themselves by war. 2. Having as a tribe given up war since we settled among them, and having as a consequence become poor, they wished to know how we were to enrich them, as they expected us to do if they gave up war.   3. That all the members of the Mission were to leave Bandawe and come to Ngoniland, so that instead of there being only one station the whole country might be occupied.

The questions were formidable enough, and it now became apparent to us that we would have to deal with the clamourings of Mombera's brothers who were chiefs of large districts. We were thrown into great anxiety, as there seemed to be no way out of the difficulty. The first question had always been a source of trouble to us, and as the years went on their jealousy of the Tonga increased, because they considered them as more favoured by having the head station among them, and imagined that they received unlimited supplies of cloth and beads. The second was likewise present always, but its prominence was waning, as our work had greatly turned it out of the minds of Mombera and his own retainers, but to have to begin to fight the other chiefs on this ground filled us with great fear for the safety of our work even among Mombera's people. The third question showed that it was not missionaries but calico-distributors that Mombera's brothers desired. The reason of their desire to have all the missionaries in Ngoniland was, that they might retain us and yet have freedom to wage war on the surrounding tribes.

The situation was at least clear to us, and we could set about making our arrangements. As the continuance of the work at Bandawe depended on the attitude of the Ngoni towards the Tonga, we in Ngoniland had to act in concert with the brethren there. Our dilemma was this : Ngoniland could be held by agreeing to Ngoni demands, but that involved casting away the Tonga and leaving them to the inhuman attacks of the Ngoni. They had helped in opening up Ngoniland—some of them even losing their lives in our service—therefore on no account could we think of that. But if Ngoniland could not be held, neither could Bandawe, and the Tonga and we together would suffer. Yet to accept the Ngoni proposals would have been to take sides with them against the Tonga, and not for a moment did we think of doing so.

I wrote and urged Dr Laws to come up and meet the Ngoni with George Williams (the Kafir Missionary) and myself. At this time we suffered under a heavy family affliction, and my wife was lying helpless in bed. Around us were the Ngoni in a very unsettled state, engaging in war-dances every day. Below the house near the river the armies of Mtwaro and Maurau, Mombera's brothers, were encamped, bent on some expedition, the nature of which was hid from us. To complicate matters, the road to Bandawe was closed, and carriers could not be got to go down. It was a time of terrible suspense, and although not of personal danger we believed, the fear that our work among the Ngoni and at Bandawe might be ruined, filled our minds with uncertainty and distress. It seemed the darkest hour of our life among the Ngoni, and our neighbours were afraid to be on intimate terms with us.

When letters passed between us Dr Laws and I decided that as there did not seem to be much prospect of a peaceable settlement, we should endeavour to prepare at both places for being driven out. Our situation in Ngoniland was anything but pleasant or easy. Our letter-carriers were the two slaves before referred to, and in their journeys to and from the Lake they left the usual paths and travelled in the bush. They were trusty fellows and were the only ones in whom we could confide. They received their letters or loads at night and started off, getting well into the bush before daylight.

It was evident to us that in our possible expulsion from Ngoniland we should be unable to take anything with us, so we set about the saving of the most valuable of the Mission property. It is fortunate that on such occasions a missionary's own possessions do not usually stand very much in the way, and I had only my microscope and books to be a great care at the time. We could not send away many things, so our first care was to get the valuable surgical instruments sent off. We could only send small parcels as our carriers had rough ground to traverse, without paths. Over £100 worth of instruments were quietly despatched, and then my microscope followed. Everything else had to be disposed of otherwise or left in the house. My books I put in tin-lined boxes and buried in the ground. Other things were treated in the same way. The well-stocked dispensary presented a difficulty. I could not close it or have the shelves emptied lest suspicion should be aroused. I chose several spots in the garden and in one of the stores, and buried the medicines. It was the height of the dry season and the ground was hard as stone. I went out for several nights about one o'clock, and by means of an auger bored the ground under cover of darkness and scooped it out with my hands, not daring to use any tool lest the sound should attract any one who might be out of doors. We could not trust even our house boys.

All the while that this was going on, my wife was lying weak and helpless in bed, no doubt greatly hindered in her convalescence by the anxieties of the hour. For several hours every night I dug up the earth and made pits in which to bury the medicines, anon running in to pass a few minutes with my wife in her weakness. It was easy work secreting the stoppered bottles. I knew the labels would be destroyed by white ants, so to preserve the names of the drugs in the several bottles, I scratched a number on each bottle and carefully noted it in a book. For corked bottles, jars, etc., which could not be buried, I adopted the plan of putting them into an empty flour-tin, and soldering it up. In this way nearly all the drugs were preserved against the worst which we feared, and a plan of the station was made, and the spots where the things were hid, carefully marked. A copy was sent to Bandawe and to one of the other stations, and I pocketed one to carry with me. In digging in the store I had to lift a brick floor. I could only work at this when the servants were out of the house, and they had a high time of play for some days, as they were granted an unusual amount of leave. The earth taken out had to be carried away at night. There was a sense of relief when so much of the Mission property was made secure, although at the time the air was filled with the sound of war-dances, and armed parties were collecting at the chief's village for review.

It was an occasion requiring great prudence, and we had to act in Ngoniland, not only because of our own position, but to give Dr Laws time to arrange matters at Bandawe. Dr Laws was aware that there would probably be trouble with the Tonga. He found that, as on former occasions, they demanded the assistance of the white men in meeting the Ngoni, and some even proposed that they should accompany them in a Tonga invasion of Ngoniland. In the middle of September Dr Laws gathered the Tonga together, and put before them the whole situation, as it came out in the questions the Ngoni proposed for discussion. They declined to give up the women and children whom the Ngoni claimed, as the Ngoni had stolen them in the first instance. By the Lake steamer which had brought reinforcements for the Mission, some goods were sent away from Bandawe, and also native women and children belonging to Cape Maclear. The Ngoni were clamouring for Dr Laws, and we had to put off fixing a date for the meeting until he could see his way to leave his fellow-members safe at Bandawe. We had to try and arrange so that the flight of the Bandawe staff should be possible in the event of war, leaving us on the hills, with Dr Laws and the faithful Tonga he was to bring with him, free to act according to circumstances.

All seemed to go well for a time. We had managed to get the more valuable Mission property away to Bandawe, and Dr Laws had sent that and also much of the Bandawe goods away to Cape Maclear, and the steamer had returned to stand by and take away the rest of the staff if need be. But just as Dr Laws saw all this arranged and was ready to come to Ngoniland, his difficulties at Bandawe were suddenly and greatly increased, and we had again to postpone the date of meeting. The Tonga had become suspicious of the movements made at Bandawe, and prevented the embarking of passengers or goods on the steamer. The roads were watched day and night lest an attempt should be made by any one to go away, as they feared that if once the white men left Bandawe they would be destroyed by the Ngoni. It was therefore impossible for Dr Laws to come up before the end of October, and even then he had the anxiety of knowing that the fickle Tonga were surrounding the station, and that not one but all at Bandawe and in Ngoniland were now forced to await on the spot whatever the issue might be. We knew from the first that we in

Ngoniland had to wait, but it was not expected that the Tonga would rise as they did and thwart the best efforts of Dr Laws who, with us, was enduring all the trouble on their account.   As he wrote to me at the time, " The matter comes to this, we must do our best to stick to both stations if it is possible for you up there to do so with safety.   If you must leave, then some other point than Bandawe must be the point aimed at on our return, for to come here would be simply to be fixed in a trap with the rest of us.   To-night I feel that here I have sought to do everything that prudence might suggest for the safety of lives and property.   Now we are hemmed in, and we can only await to see what may be the next indications of God's providence, trusting our Heavenly Father to guide us to do what is right, just and true, and altogether according to His holy will."   For a month longer the suspense had to be endured, and the letters which we were at times able to send each other are full of what was in our minds, and what was so prominent a subject in our prayers.   Each letter was closed, not knowing what might transpire before another could be sent, and gave indications of what was to be done should the Ngoni rise up.   We had to hide from the Ngoni the cause of Dr Laws' delay in coming up.   Had they known of the action of the Tonga at Bandawe they would have gone down with war. We had to trust our two faithful slave-carriers not to reveal what they saw and heard at Bandawe, and throughout it all they were faithful to the trust. At this time the Ngoni army went out, and conflicting rumours of its destination only added to our anxiety. It went, however, to attack some Arabs and Bemba towards the North-West.

At length, on the 27th October 1887, the great meeting of the chief Mombera and his headmen with us was convened. We met at eight o'clock in the morning, and did not end the *indaba* until four in the afternoon, and thus sat in the open cattle-fold under a hot tropical sun for eight hours, the discomfort to Dr Laws being very great as he was suffering from fever and had to go to bed for two days when he returned to the station. When the *indaba* began we were asked to state what we had to say, but we refused to begin, saying that they had called us and we would hear them. Mombera thereupon said that all the missionaries must come and live with them, and leave them free to attack their former slaves, the Tonga. We reasoned with them about the necessity of having a station at the Port of Bandawe in order even to carry on the work among the Ngoni on the hills. They conceded the point, but when it came to the question of fighting the Tonga, we had to spend hours of talk ere we got the concession that no war should be carried to the neighbourhood of our station at Bandawe. The embassy from Mtwaro was not agreeable to any cessation of war, and at length declared that they at their end of the country never saw us or received anything from us, and they were determined not to submit to stay at home while we went up and down to the Lake. Our road, they said, did not lead past them and there were no crumbs for them. We seized upon this remark and offered to open a station in Mtwaro's district, which satisfied them. We could have told them that we had tried for years to enter their region but were hindered by Mombera; we refrained in the circumstances.

The only man that day who could not be won over was Mombera, but as it was a tribal matter the decision of the councillors was binding on him and the matter ended. The story is soon told, but as the day went on the demands of the Ngoni were crushing to our hopes of peace, and we had even begun between ourselves to formulate a plan for saving Ngoniland and Bandawe by Dr Laws himself exchanging stations with me for a time. As Dr Laws wrote at the time, he " well knew the worth of their request to have him, that it was simply inspired by the desire to squeeze as much calico, etc. out of us as possible." But from the quarter whence came the heaviest assault that day — Mtwaro — the solution came also, when his ambassador desired to have a station in their district. With light hearts we went home and soon messengers were speeding through the forest, over hill and mountain-torrent, bearing the glad news to our friends who had meanwhile been left in uncertainty at Bandawe, and hearty were our praises that night at worship with the natives. The Tonga who had come up with Dr Laws had been confined to the station, but now they freely and joyfully mixed with our Ngoni neighbours. Peace had been declared and at no time since has it been broken. On one occasion years after, when some wild youths went on a marauding expedition to the neighbourhood of Bandawe, Mombera, the chief, called them up. He said, "You are not chief. I am chief. You went to Bandawe with war. Cut their legs," and they were thereupon hamstrung. "You killed Tonga. Cut their wrists," and the tendons were divided, and the miserable wretches crawled away to hide and die.

We had also received an invitation to visit Mtwaro, between whom and the chief Mombera, his brother, there was a feud at that time. Each blamed the other, as was referred to in a former chapter, and while under Mombera we were not permitted to visit Mtwaro. When, a few days after the *indaba*, Mahaluli, Mtwaro's ambassador, visited us, we desired him to settle with the chief that we be permitted to accept the invitation. He did so, and we lost no time in going to Mtwaro. Our way was plainer than on a former occasion, when we went to treat a child of Mtwaro's, who was sick. We received a hearty welcome, and when the boy who had been sick was brought out and proudly shown as now in good health through the white man's medicine, it was evident that the effects of the medical work were wider than in the good recovery of the lad. Some days were spent at Mtwaro's, and a frank invitation to remain among them was addressed to us. They had heard of the medical work, and they wanted a medical man. They had heard that we were rich, and would give them much cloth. When we spoke about our work, they were not sure if they should give their children to be taught. The old fear that they would be bewitched by our words, and would not engage in war, was expressed. They had heard of the Book, and what it had done, but they thought we ourselves should " practise the Book," and give rain and good crops and success in battle, but leave the children alone. We were prepared to hear such things, and illustrated how they should think otherwise. We had with us Njuyu boys, who could read and write, and we astonished them by showing what could be done by means of writing. A boy was sent out, and then the chief dictated certain words, which I wrote down, and the boy on being re-called read them. I was then sent out, and things were hidden, and I was requested to say what they were and by whom secreted, which I did as the boy had written them down. But it appeared to them as if it were only the magic of their witch - doctors, and they thought that it would be unsafe to let all the children into the secret. When we came to talk about God and His Word some of the old men left, being afraid; but before the visit ended we were permitted to conduct a service, and in the singing they were specially interested. We were offered a site for a house and garden, and requested to come and build.

Another visit of importance was thereafter paid to Ng'onomo, the Prime Minister of Mombera. He was a very old friend, but was the leader of a large army, and frequently out raiding. He invited myself and my wife to "come one day, stay two days, and leave on the fourth day." We were received in a friendly manner, and as a Sunday intervened we held a service. He did not attend, but ordered his people to do so, and we had a large gathering near our tent. It was the first Christian service ever held in the district, and, for uproarious behaviour of the audience, was never surpassed anywhere. They roared with laughter when we and our Njuyu men closed our eyes in prayer. Some men who had been at an Njuyu service understood our object, and tried to quiet the people. That was not easily done, and they fell to cursing and swearing while the prayer went on. An address followed, which was listened to with some attention. The chief, Mombera, arrived while we were there with a large retinue, to pay Ng'onomo a visit. Cattle were killed, and large quantities of beer were brought in from other villages. The Sunday was spent in riot, and the only quiet we could get was by going out to the bush and spending some of our time there. We were close to the Hora Mountain, and the ground was strewn with human skeletons—the remains of the poor Tumbuka who were slaughtered some years before by the Ngoni. The arrival of the chief, and the debauchery which ensued, prevented all serious talk with Ng'onomo, and we were glad when we could leave the place and return home.

When we again settled down at Njuyu the work, which had all been interrupted, was resumed, and the school and services at Chinyera, five miles distant, were begun again by Mr Williams. We also set about transferring our medicines from their place in the garden to the dispensary shelves, and under cover of darkness that was done. They were uninjured by being buried, and we restored the labels. Books suffered, and certain other things buried in insecurely-closed boxes; but we were too overjoyed at the safety of the work to be pained by our losses. In a short time all was as before, and we had our hands full.

The rainy season of 1887-8 passed without trouble to us, and during the next two years we greatly extended our work of preaching and healing as we were free to move about almost everywhere. We were unable, for want of a man, to open the station at Mtwaro's till near the end of 1889, when Mr and Mrs McCallum, from Bandawe, began work there, and conserved the results of our frequent visits and operations there in the interval, proving earnest workers in a field where many difficulties were present. Our attempt to gain access to a large district under another brother of Mombera, named Mperembe, proved futile, as we were repulsed on several occasions because we would not pay our way by presents of cloth and other goods.

For a few months in 1889 I had the assistance of a European at Njuyu, and school-work flourished. On his retiring, Mr Charles Stuart, who had newly arrived to reinforce the staff, was located at Njuyu, and had to take up the whole work. As in the end of January 1890, I had to go to Bandawe to relieve Dr Laws who was ill, and thereafter in May to start for home on furlough, which was considerably overdue. Ere leaving, however, I was privileged to see the accessions to our staff; the baptism of our first converts, Mawalera and Makara, two of those who came by night to be taught in the dark days of our history; the extension of the work to Chinyera and Ekwendeni, and the institution of five schools. A beginning had been made, and the long, weary years of waiting crowned with liberty to go about and "heal the sick, . . . and say unto them, The Kingdom of God is come nigh unto you." Our fallen wattle-and-daub school was replaced by a brick building, which in a few months proved too small to accommodate the scholars, and was replaced by a large, brick school, which in its turn had to be enlarged.

DR STEELE.

CHAPTER XII

FROM 1890 to 1895 the work in Ngoniland was superintended by Dr Steele, who was appointed to relieve me on my departure on furlough. His arrival, as the previous chapter indicates, was at a time when a distinct stage in the work had been passed and a new era begun. The formal consent of the chief and most of his head-men had been obtained, and advantage of the opportunities offered had been taken. The nature of our work had become more apparent, and it had begun to bear distinct fruit. Death had come and claimed two who fought bravely in the early battles. George Williams had resigned and returned to the Colony, and two additional white men had been initiated in the work, so that with increased and more earnest attendance at school and service in three distinct districts, with a roll of two Church members, the position of the work on his arrival was full of rich promise. It is a long story to confine to the limits of a single chapter, but the tale of the work during those years may fittingly enough be associated with the name of our dear departed fellow-worker, whose death took place at Ekwendeni on June 26th, 1895, when he was on the point of going home for his first furlough.

Dr Steele was the youngest of seven children, and when only two years old was deprived of both father and mother within the space of three months. The eldest was only seventeen, and she and the older of the brothers determined that, however dark were their prospects, they would endeavour to keep the family together, and take upon themselves the care and support of the younger members. At the age of eight he entered the Buchanan Institution in Glasgow where he remained until he was thirteen, receiving a good elementary education there which proved a sound foundation on which, by his own ardent efforts, he ultimately proceeded to academic distinction in Glasgow University. On leaving school he entered a drapery establishment, where for some years he remained and carried out his work with his naturally exact and painstaking fidelity. At this time his brother Richard (next in age to Mary the eldest) had commenced business in the boot and shoe line, and Tom, another brother, assisted him.   Tom, however, determined to study for the ministry.   During Mr Moody's first visit to Glasgow, he was one of the hundred young men who offered themselves for service in the Lord's cause, and for whom arrangements were made for special evening classes to enable them to carry out their intentions.   Tom eventually passed into the University, giving up his situation with his brother, and George left his work in the drapery establishment to take Tom's place. The boot and shoe business supplied but a very moderate income, on which the family depended, but indomitable perseverance and a strong family affection kept them together. Tom at length passed through the University and the Divinity Hall, and is now a licensed minister in the Colonial field.

When George was twenty years of age his future became a matter of concern to him.   He had given himself to the Lord in heart, and now he desired to give his life to His service.   His brother Richard encouraged and helped him, and so he began to prepare for entering College with a view to become a medical missionary.   He attended evening classes for four years, and continued to work under his brother in the shop.

His progress at evening school was good, because he studied faithfully, and took from the hours for sleep what time his attention to business deprived him of. At length he entered the University, and it was only then that the strain of the financial struggle began. His brother Richard had given up his business and had followed Tom to Australia, and now George was thrown entirely on his own resources. It is an oft-told tale of hard work, high ideals and aspirations, overcoming in great trials. But by rigid economy practised in all details of his life, without meanness or disregard for the position he occupied, and for which he had to commend himself by habit and personal appearance, and above all with an implicit faith in God, he passed through the medical curriculum and graduated, the possessor of the coveted degrees in Medicine and Surgery. The same story might be told of many in our Scottish Colleges, and the case of Livingstone was no doubt a stimulus to him who was to be one of his direct successors in the great work of the emancipation of Africa and its ingathering to Christ.

When Dr Steele was attending Glasgow University he was one of a band of noble young men who have, like him, entered the field as medical missionaries. It is something to know a man's friends, especially his College companions, and I need only refer to Rev. Drs Mowat, Sandilands, Macphail and Revie, all in the India Mission field. Before entering on the record of Dr Steele's work in Livingstonia, I have pleasure in giving the tribute of Dr Macphail to the memory of his companion and friend.

" It was my privilege, from 1884 to 1889, to have the late George Steele as a class-fellow attending the medical classes at Glasgow University, and it is a pleasure to have an opportunity of adding my tribute to his worth. In the year 1884 the religious movement among students, identified with the name of Professor Henry Drummond, spread from Edinburgh to Glasgow, and was the means not only of leading many of the students to devote themselves to the service of Christ, but also of bringing those who had already given themselves to it into closer fellowship with each other. It was at this time that George Steele became intimately known to me; it was the beginning of a friendship which we maintained in after years by correspondence when he was in Central Africa and I in India, and which was only brought to a close by his lamented death. He threw himself heart and soul into this movement, willingly lent his aid to the work, and also joined a small society, which consisted of students who had dedicated themselves to foreign mission work. He was a very diligent, hard-working student; and like so many others in our Scottish Universities, had to maintain an incessant struggle with adverse circumstances while pursuing his studies. But this did not prevent him either from retaining through it all a very happy-hearted, cheerful disposition, or from taking a large share in the work of our Christian societies. For a season he was President of our Foreign Mission Students' Union—afterwards amalgamated with the Volunteer Union—and many of us remember with pleasure the papers he contributed to our meetings. Towards the end of his College course he was appointed assistant at the South-side Medical Mission, and soon earned the grati-tude and affection of the poor people to whom he ministered his kindness and skill. At the same time the thoroughness of his class-work was indicated by the fact that he won the medal in session 1887-8 in the Junior Division of Professor M'Call Anderson's class of clinical medicine—an honour which brought almost as much gratification to all of us who knew him as to himself. It was some recognition of the conscientious, painstaking spirit which characterised him in all he did, and was all the more creditable to him because he was one of those who worked not for academic distinction but to qualify himself for the service of Christ and humanity.

"We spent a short holiday together in Arran before I sailed for India in 1889, and in the days of recreation one learned to love and admire him as much as in the hours of labour. His nature was singularly bright and buoyant, and his keen interest in botany and natural history added greatly to his enjoyment of country life.

"Like most of those who have been led to serve the Master in the foreign field, he was deeply interested in the work of Christ at home. He was always a loyal son of the Free Church of Scotland, and was warmly attached to Free St James' Church in Glasgow, a congregation which showed their appreciation of him by contributing largely to his surgical equipment when he left for Livingstonia. Of his work in Africa he often wrote to me in terms which showed how much he loved it. His brief sketch of camp life among the Ngoni I read with very great interest.

"It was with very real sorrow that I heard of his death. Neither time nor distance had lessened our friendship, or my regard for one whose memory I shall always cherish as one of my most treasured possessions."

My first meeting with Dr Steele was on board the *Courland* in June 1890, in the Quilimane river. His appointment to Livingstonia had been made known six months before, but circumstances had rendered my departure necessary before he arrived. When the steamer arrived I went on board, and found him with Messrs James and George Aitken for Livingstonia, and six young men for the African Lakes' Company. All were full of life and hope, and yet only two of the party were spared in health to return at the end of five years. Such is the sad side of life in tropical Africa! I spent several days with Dr Steele, and saw him away in the boat for his first and last journey up the Kwakwa river. He entered Africa full of buoyant hope, and during those few days, from the questions he put and the views he expressed, I was greatly impressed with his suitability for the work among the Ngoni to which he was appointed.

His introduction to African life was of an unusual description. He was at Quilimane at the time of the Anglo-Portuguese difficulties over the opening of the Zambezi and Shire rivers, and the passage up the latter was made at one point through a shower of bullets from the Portuguese guns at the Ruo. The late Joseph Thomson, the African traveller, was with them on the steamer, and his experience and ability got them safely beyond these opposing Portuguese. Dr Steele's letters, written on the passage out and during the inland journey, are full of beautifully simple and graphic descriptions of all that he saw and passed through, but running through all is the index of his thoughts of the work he was to take up, and of the moral and spiritual condition of the tribes he passed through. Nothing escaped his notice, and all being so new, he wrote fully and in an interesting manner on subjects political, scientific and spiritual. Reading through the pile of his letters before me, one is struck by the self-effacement shown, and his surprise at certain things which disproved his preconceived notions of men and things as they are in Africa, but yet the openness of mind with which he listened to those whose wisdom and experience fitted them to be his helpers. Ardour, humility, hopefulness of disposition, and consecration to God shone out in his life from the very beginning.

Dr Steele reached Bandawe in the end of July and settled at Njuyu in the beginning of August. From the time I left in January to relieve Dr Laws at Bandawe, until his arrival, the station at Njuyu was held by Mr Stuart, and that of Ekwendeni by Mr and Mrs McCallum. These lonely workers, fifteen miles apart, among a troublesome people, were new to the district. Mr Stuart had but a few months' African service, while Mr McCallum had come from his former work at Bandawe. At Njuyu especially the lot of Mr Stuart was not an easy one. Surrounded by the discontented Chipatulas, whose ambitious desire for power and wealth through the aid they expected from the white men at the beginning of the Mission had not been realised, he was subjected to many indignities as they took advantage of his novitiate. At Ekwendeni Mr and Mrs McCallum were in the throes of instructing the lazy Ngoni in manual labour and in education, and had in that new district to meet and overcome the begging habits, proud arrogance, and dark superstition of the people. Only those who have had experience in the beginning of such work can know what it means. The arrival of Dr Steele cheered the solitary and hard-pressed workers.

His first visit to Mombera was much a dread to him, as every account he had received of how new-comers were treated had made him shrink for some weeks from facing the ordeal. Then Mombera was in one of his bad moods and the reception was not very cordial. When the subject of the school, which he had said before he desired at his head village, was mentioned, Mombera would not sanction it. His first visit to Mtwaro at Ekwendeni was more cordial and pleasant, and Dr Steele's offer of help in the treatment of his diseased joint drew Mtwaro very near to him, and compensated for the rudeness of Mombera—a rudeness which few who saw him often do not understand.

As they had been some months without a doctor, the arrival of Dr Steele was hailed with delight by the people. Medical work attaches a people more quickly than any other. While they could not understand much of what was taught, and often did not realise that it applied to them at all, all who received help in distress, and relief from pain and disease understood that, and counted the Mission-doctor their friend. Very soon however Dr Steele realised that what they desired most was wealth, and their begging habits were a continual annoyance to him in his work. Here is a graphic picture of a common occurrence in the course of a visit to the villages in carrying on medical mission work. He says, " The first place I stopped at was Zigodo's village. This man is Mtwaro's head councillor. To excite my pity he appeared quite naked, but two yards of cloth sent him away rejoicing. My next place of call was Sunduswayo's where I dressed his mother's ulcer.   He told me I was to come again soon and bring him cloth.   I next called at Hlojana's.   In conversation with him about a pain in his shoulder I asked if I might bring more medicine for it.   He replied that I was to stop with him next time I passed and give him a piece of cloth.   Few head-men are above begging."

Reflecting on his work and surroundings as he wrote to a friend, he drew a vivid picture of the condition of the people.   "Up here, shut off from the rest of the world by the eternal hills, they have little or no notion of what is going on in the great universe outside.   The white men come from somewhere, but where, they cannot tell.   How to bring home to the natives the Gospel we have come to teach is the problem requiring much earnest thought, and above all, heavenly guidance.   Their very simplicity of mind makes it difficult, but this condition of mind has combined with it the vices of men, so that while intellectually they are children, morally they are men, for one sees among them all the vices of human nature,—pride, avarice, greed, meanness, dishonesty, falsehood, &c.   Despite these, however, they are not destitute of good parts."

Dr Steele was not long among the Ngoni until

he discovered, as others had done, that they have a standard of manhood all their own. Until a man married and owned cattle he was considered to be but a child and had no interest in tribal affairs. This made the position of the unmarried on the staff very liable to be slighted, and shut doors of usefulness against them. Dr Steele was considered, like Mr Stuart, to be a mere boy and not to be admitted to serious conclave along with men. He recounts that "one day while Mr Stuart was engaged building the school a proud Ngoni addressed him as a child, contemptuously. ' Yes,' he replied, ' I am a child as you suppose, but can your children build houses like these ?' The man called him no more a child but was very respectful, and it seemed to dawn on his mind that manhood did not consist in the possession of wives and cattle but in knowledge and wisdom."

As I can testify, the arrogance of the Ngoni was at times hard to endure, but an incident such as the following served to sweeten life in his case as it had done in my own. Dr Steele writes, " About four weeks ago I was called to one of the chief's villages to a complicated labour case. All went well. The gratitude of the people seemed very sincere. The women came round me on their knees, slowly clapping their hands and saying in a tone of great relief, 'my father,' a term of great respect. One woman rolled herself on the ground at my feet, but I could stand this no longer, and told her to rise up. I told them all sitting round that God sent me to heal the sick and tell them about Him. I then doctored some twenty eye-cases and returned home." When it is mentioned that such women as needed assistance in maternity were accused of foul crimes and usually consigned to the bush to die, and their friends taken as slaves by the husband, the gratitude of the women will be known to have been sincere. No branch of medical science has been more fruitful of life-saving and of good to the helpless than that, and all medical missionaries have found it so. The glimpse we get of medical mission work in Dr Steele's hands explains the attachment of the people to him.

A time of great anxiety came upon Dr Steele and his co-workers, Mr and Mrs MᶜCallum, at Ekwendeni, when Mtwaro the chief, who was under treatment, died. He suffered from chronic disease of the knee joint, which I had been called on to treat a year before. It was evident even then that nothing but amputation could be of use in saving his life, but as he did not submit, palliative treatment was adopted. When Dr Steele

arrived he was warned of the case, and advised
not to adopt any active treatment until he became known to the people and had their confidence, for, on account of their superstition, any
untoward result would be blamed against the
doctor, so the people, and not the patient only,
had to be considered. They could not realise the
seriousness of the case or understand the treatment adopted, and might be incited by the
native doctors to rise against him, and so the
work be hindered. Dr Steele's treatment was
only palliative, and as it turned out, a rapid
extension of the inflammation caused death in
a few weeks. During those weeks, when the
illness of the chief made the people very excited, the cry went forth that the white man's
medicine had killed the chief. Various divining
doctors were called, the poison ordeal was gone
through with dogs and fowls, but again and
again the verdict was in favour of the doctor.
The people were, however, very excited, and in
the event of death it seemed as if trouble would
come to the Mission party. Mr and Mrs McCallum
were there alone, except when Dr Steele was with
them in connection with the case, and the occasion
was one of anxiety such as they only, who have
witnessed a superstitious people driven frantic by
some event, can understand. Mtwaro died, but

while he was conscious he declared his confidence in the white men, and sought to quiet the people. For a time all work was stopped in the district, but the people quietened down and the work was resumed again, the deceased chief's widows and children coming freely to the services and school on the station. In those days the medical man had to contend with superstition, and with the combination of divining and medicine men who witnessed the progress of the work and emancipation of the people from their clutches, and were therefore too often opposed to the Mission.

In the middle of 1891, Mr Stuart, the teacher who was with Dr Steele at Njuyu, was transferred to fill the breach caused by the invaliding of Rev. Dr Henry at Livlezi station among the southern Ngoni. In Dr Steele's letters at that time, he writes joyfully of the extension of the preaching work into new districts; of his three schools and twenty teachers, and of the services and Bible classes. At the close of his first year he wrote thus: "In the course of another week I will have come to the end of my first year. It has been full of many new experiences, and I must say has appeared long. I am beginning now more vividly to realise that this is now my home than ever I did before. The reason is that until one gets a hold of the language you feel estranged, but when you begin to understand them better and speak their language, the feeling of separation begins to pass off, and one feels, as it were, one with them." These words evidence how truly he was ripening for the succeeding four years of full, patient and productive service, ere he was called to the service above.

When Mr Stuart left for Livlezi, his place at Njuyu was taken by Mr Scott who had just joined the Mission. In consequence of the change, Dr Steele had all the work of the station laid on him for a time. He had the hours of every day well filled up by medical, evangelistic and educational work, and had also, as so many of all classes in Livingstonia have had to do, to lend a hand in the brick-field, and at house and school building. The work was in full swing when an event happened, which a few years before would have probably been fraught with disaster to the work. Mombera, the chief of Ngoniland, died suddenly. The death and burial are described so graphically by Dr Steele that his account of it may be given here as affording a picture of African customs in such circumstances.

"Mombera died somewhat suddenly. I saw him the Sabbath before, when at his village conducting a service. He has been chief of this section of the Ngoni for many years, and his death has therefore caused a great sensation throughout the country. Poor Mombera! If he had been inclined to know the Gospel, he has had good opportunity for years, but he never showed any interest in it, and died as he had lived—a heathen.

"The people came from all parts to take part in the mourning ceremonies. He died in Empikisweni and his body was carried to his own village, Engaraweni, where he was buried. We left the station early in the morning and arrived at the village about noon. On nearing the village, many people were seen coming in companies to mourn. The men all carried shields and spears, and some had also guns. As is the custom, the people had removed all their ornaments, and some had bands of grass tied round their heads. When we entered the village, the men who were with me stood together, and raising their shields over their heads, began to utter a wild continuous cry, 'Baba be! Baba be!' *i.e.* 'My father! My father!' The cry was taken up by each new company and continued for a time when they retired to make room for others. After this we went to the public place — the cattle - fold — which was crowded with men. There we waited till dark.

The continual wail and all our surroundings produced a strange effect on us.

" When they had consulted as to the grave they set to work to dig it. At this stage a sister of the late chief appeared at the grave and began to mourn. Suddenly Mperembe, another Ngoni chief and brother of the deceased, sprang up and began to mourn. He placed his hands clasped behind his head and advanced with a dancing motion to the grave. All the men in the cattle-fold did the same. This went on till nightfall when the work of digging the grave was suspended.

" Next day about ten o'clock the grave was dug and preparations were made for the interment. All the morning, however, the late chief's wives were engaging in a ceremonial. Some of them carried the chief's shields in front of them and approached the grave mourning loudly and retiring again. All the old women took a prominent place in the ceremonies. Before the burial the cattle-fold was cleared of all except married men. Just before the corpse was brought in, the chief's wives formed a strange procession. They crawled up to the grave on their hands and knees in single file, mourned for a short time, and then retired. They were all dressed in skins and had their heads covered with cloth and great bunches of feathers.

" Some men now left to bring in the corpse. They went towards the hut with their hands clasped behind their heads and wailing loudly. They returned bearing the corpse, which was followed by a large number of women carrying things belonging to the late chief to be put into the grave with him. The men stood in circles round the grave with their shields high above their heads and wailing piteously. This continued for a quarter of an hour, and then they gave place to the numerous companies of young men who at this point came in to mourn. Company after company filed in, and for a long time this went on. After the body was placed in the grave the chief's pipes, pillows, etc., were put into the grave. One grave was completely filled and another had to be dug into which the remainder of the chief's personal belongings were put."

The furlough of Mr McCallum in 1892 and my return to be at Bandawe necessitated further changes in the staff. Mr Stuart relieved Mr McCallum at Ekwendeni ; and in view of future developments of the work, Dr Steele, accompanied by the late Dr Henry, made an extended journey round the country so as to find out in what direction our developments should be made.

The work at the various stations was now fully organised in open-air and school-services, day-schools, and Bible-classes for men and women, and its effects were seen in the changed habits of many around the stations. At Njuyu under Mr Scott a great advance had been made in school-work, and singing had been greatly improved by the introduction of a harmonium. The elderly women's class, begun by Mrs Elmslie years before, was being better attended, and the time spent in it was bearing good fruit—the women not only being eager to be taught the Word more systematically, but evincing a willingness to be taught to read and write. At Ekwendeni, not only had the principal people of the district been reached, but a mighty advance had been made through Mr McCallum's school, in that master and slave had been brought together on the same level, and the first natives on the staff there belonged to both classes.

In 1892 Dr Steele had the joy of baptising the first Ngoni woman, who was the fruit of the adult women's class referred to, and eight men. The church roll now numbered eleven adults and four children. Two years before, the first converts were baptised. But still there was much ungodliness around the stations, and gross darkness. The dry season brought together great companies of people to drink beer, and then, as Dr Steele wrote, war was a subject much talked of, and several bands of reckless young men went out, among whom were some who were attending school and Bible class. He wrote at this time, "But some people from whom I expected better things are still remaining very superstitious, and are inclined to say hard things about us, and to treat badly those who believe our message. Lately one of our people was at one of the late chief's villages some four miles from here. He was maltreated because he affirmed that the dead would rise again. One man got so exasperated that he rose and left the hut. They still harbour silly notions about us, such as that I killed Mombera, the chief. They call me an *Umtakati, i.e.* a witch, a villain who can harm people by bewitching them. Many of the people are still inclined for war, and this along with their superstition and ignorance makes them disinclined to receive our message, and they become jealous and envious of those who have believed and are perhaps getting in advance of them. Some speak well of us and of our work, and others cast reflections on them for doing so, and out of jealousy and spite they defame us.

For example, the second wife of the late chief, of whom I had good hope, has disappointed me, and because the third wife has been bravely defending us the second has brought a charge against her of having killed the chief. We must have patience and live these things down."

In 1893 Mr and Mrs M‘Callum returned from Scotland, and Dr Steele was enabled to open a new station at Hora by locating them there. This is the district of Mzukuzuku, a famous general in the old Ngoni army, and the one where the great massacre of Tumbuka took place about 1880. For the second time Mr and Mrs M‘Callum had to begin the arduous task of introducing the Gospel to a new region, and although each successive year saw the whole tribe more amenable to the Gospel, the work was trying enough. As we now witness at this date, the pioneering both at Ekwendeni and at Hora was successfully accomplished, and the developments now visible are in great measure due to the work of these two missionaries at the commencement. As recorder of the history of the work it is my duty and pleasure to state that.

There were now three stations manned by Europeans. Comfortable brick houses were erected at each, and, as the country is healthier than other parts of Livingstonia, the work was vigorously pushed forward. In 1894 the number of schools had increased to nine; special instruction was being given to over thirty native teachers who were all under indenture for five years. The Bible classes increased in number and attendance, and the village services were being conducted by Christian natives as well as by Europeans. While this new district at Hora had been opened to the work, an attempt which was made to get into that of Mpcrembe, a brother of the late chief, failed as it did on a former occasion. But Dr Steele and his colleagues had learned that persistency of effort in districts open, and frequent calls upon outside parties, would eventually win the way for the Gospel, and the whole history of the work showed that great patience was necessary. No district could be taken by a rush, and so what was gained was held by patient and persistent effort.

Here is a picture of a day's work. " You will rejoice to know that the new bell is now erected and doing duty. I don't spare it. Since it was erected I have commenced a morning service. It is rung at sunrise and we all turn into school, *i.e.* Mr Scott and myself and our eight boys, teachers, workers, and the general public. First a hymn is sung; then I tell the day of the month and the text for the day which all repeat after me several times. Selected portions of Scripture are then read and all join in the Lord's Prayer, after which each goes to his work. Again at half-past seven the bell is rung for the school and our breakfast. School commences at eight, and this junior school is conducted entirely by natives under Mr Scott's superintendence. The junior school is over at half-past nine, and the senior school for the teachers which, till twelve o'clock, Mr Scott and I conduct, begins. We rest till two, and on Monday, Tuesday, and Thursday, I have my women's school from two till four; on Wednesday and Friday the catechumens, and at five o'clock all public work is over for the day. On Wednesday evenings there is a prayer meeting." The same order practically obtained at Hora and Ekwendeni, special attention being paid to teaching the teachers, not only for school-work but with a view to qualify them also to become evangelists, such work being put into their hands whenever their consistency of character and attainments made that advisable. In the beginning of 1894 there were 760 children attending the schools in the three districts.

In 1894, as the staff had been reinforced by the

arrival of the Rev. Messrs Dewar and MacAlpine, and the settlement of the latter at Bandawe, I was free to return to my former field in Ngoniland. My place, as indicated, had been ably filled by Dr Steele, who was in charge of the whole district. At my request he continued to occupy my former station and to act as missionary-in-charge of the district until, a year thence, he should leave for home. I settled at Ekwendeni where the first temporary house built by Mr M°Callum had been replaced by a substantial brick building erected by Messrs Stuart and Murray, and relieved Mr Stuart who left on furlough after five years splendid service. We were then in still closer contact through our work in Ngoniland, and had frequent intercourse together. He again had the joy of receiving seven more adults into the Church by baptism at Njuyu. It was the station where the work had been carried on for the longest period, and the fresh baptisms were an occasion of interest far beyond, and reacted on the other stations, so that early in 1895 we find the Church roll increased to 59 by further baptisms at Njuyu and Ekwendeni, while at Hora, the youngest of our stations, several had already made a profession of faith and been admitted as catechumens.

On my settlement at Ekwendeni we originated

quarterly conferences of the workers, meeting at
the different stations in rotation. These were
occasions not only of great good to the members,
but the plans of work, reports of progress, and
discussion of difficulties, engaging our meetings,
bound us very closely together, and strengthened
our united efforts among the heathen. By having with us Christian natives, the Church members
in the different districts were brought into closer
contact, and the work made to appear one, and
the Church one. The last year of Dr Steele's
service was very full of work, and the prospects
all round were very bright. A new development
had taken place in the end of 1894 by the arrival
of Miss Stewart, the first missionary for the
women, and she began work at Ekwendeni. At
Njuyu the special work of my wife when there
had been in part continued by Mr Stuart and Dr
Steele. At Ekwendeni and then at Hora Mrs
McCallum's work had made progress among the
women, and produced good results; and now at
Ekwendeni my wife's efforts were supplemented
by Miss Stewart, and she struck out among the
wives of catechumens and converts who were not
reached by the ordinary schools. Thus the work
was going on when the last months of Dr Steele's
all too brief service had come.

In April he left to attend a Council meeting at

the newly-opened Livingstonia Training Institution. He had food with him for five days, but as it was the rainy season and as his guides led him by a new and untried path over rugged mountains, they were detained, and the journey lasted nine days. He arrived at the Institution in an exhausted state, and we believe that in this way his strength was undermined, and when the last fever came he was unable to cast off its effects.

He returned to Njuyu, and as his furlough was due, prepared to leave for home. His last three Sabbaths were spent at the three stations, when he baptised forty-one adults and several children, the results of his own and his colleagues' work. He had sent off his boxes to the Lake and was to follow them himself in a day or two, but fever struck him down, and in five days, through utter exhaustion, he fell asleep, at the age of 34, on 26th June 1895. His funeral was attended by the whole of the Ekwendeni population, as well as many who had come from Njuyu and Hora. At the open grave a service was conducted. Amid the great sorrow of his colleagues—four of whom were present—and the great gathering of natives, his remains were committed to the dust among the shady trees within sound of the music of the Lunyanga river. There was widespread

Dr Steele's Grave at Ekwendeni.

gloom in Ngoniland. I here append the tribute I wrote in our half-yearly report at the time of his death, and thus close an all too imperfect record of a noble life and a noble work.

" No one was more beloved in our Livingstonia circle. His cheery, hopeful nature surrounded one with an atmosphere of attraction to him, and the deep springs of his disposition were easily touched by distress, whether among whites or blacks.

" His five years' work in Ngoniland forms an interesting chapter in the history of the Mission. He arrived just as the cloud which long hung over the work was dissipated, and the country thrown open to the Gospel, and the first fruits had been gathered. Speedily gaining a knowledge of the speech and habits of the people, he set to work, and by close attention to schools, which were called for by the desire for instruction, he saw the agency greatly extended and made eminently successful as a nursery for the Church.

" As an ordained missionary he gave himself heartily to the work of strengthening and instructing the Church members, and of gathering in the fruits of former sowing. As an evangelist he was never happier than when preaching the old Gospel, and going from village to village with the good news. His medical work he

always looked upon as a valuable means of
bringing the people to God, and so, while maintaining a healthy professional interest in his
" cases," he worked for the main end in all he
did, and gave his best to the work. He was
welcomed everywhere, and there are few corners
in Ngoniland where his voice has not been heard
preaching the Gospel, and his gentle hand laid
on the suffering to give health and peace.

" Besides having acquired the Ngoni language
he had made considerable additions to our
knowledge of the Tumbuka tongue by compiling
a dictionary.

" Preparing to visit Australia and Scotland, but
taken to the Father's House ! We cannot understand it ! But still, as when four of his colleagues
stood by the bedside and witnessed his triumph,
we preserve the feelings which filled our souls
and give God thanks for what He enabled him
to be and to do in the work in Ngoniland."

CHAPTER XIII

RE-ARRANGEMENT OF STATIONS AND GROWTH OF THE WORK

ON the death of Dr Steele several changes which had been contemplated for some time, were carried out in the arrangement of our agencies in Ngoniland. We were under the necessity of meeting a reduction of the European staff. From the first we had proceeded upon the plan of making use of the natives themselves as far as their character and attainments would allow, and had placed several of the schools in villages around the chief stations under them. We thus prepared the way for increasing the responsibilities of such agents. Hitherto they had been under almost daily supervision of a European, but now we ventured upon placing districts instead of schools under their charge and withdrawing the Europeans. The experiment was tried with our best teacher-evangelists, and, as the sequel will show, was attended with success in all departments of work, while, as the outcome

of it, we were able even to add to the number of
our schools.

The Training Institution, under Dr Laws and
Mr James Henderson, was opened to receive
pupils in the end of 1895. We at once selected
some from the districts under natives and sent
them there for training, leaving only those who
could be satisfactorily attended to by the native
teachers, and who were required in the work
of the district. Without the opening of the
Institution it would have been impossible for us
to have carried on our stations without the usual
complement of Europeans.

Miss Stewart, who had come out as first female
teacher for the Institution, and had been temporarily located at Ekwendeni for a year, was
at that time withdrawn to the Institution. She
successfully superintended the girls' schools
during her residence. On leaving for her new
work several of the more advanced girls went
with her to complete their training; others
who were not chosen to go to the Institution
wept bitterly when Miss Stewart left. The work
among the women at Ekwendeni was again left
in my wife's hands. The requirements of the
younger girls were partly met in the junior and
senior schools, and the older women and girls
who could not attend school were taken up by

her, as also the school girls for sewing. She also gave additional instruction to those women who were coming forward as candidates for baptism. At Hora station similar work was making good progress under Mrs McCallum who had large classes composed of women of all the social grades.

The most important change was made in the staffing of Njuyu station. It had always been worked by two Europeans, and being the oldest station, and the arena of all the old battles, the work there was known to have taken firm root, and was expected to prove a suitable sphere for our experiment. Mr Stuart, after returning from Scotland, was resident there for a few months, and when Mr Scott, the teacher at Ekwendeni, left the service of the Mission to pursue his studies in Scotland, he was withdrawn to Ekwendeni. Mawalera Tembo was installed at Njuyu to carry on the work of that district.

Mawalera Tembo has been a faithful worker for many years. He was one of those who came to our house at Njuyu under cover of night to be taught, when we were in the early struggles of the work, as has been already referred to. He and his brother Makara were the first to be baptised in 1890. His father was a witch-doctor, and Mawalera in early years had to attend him in the ceremonies of his practice, and become "art and part," consciously or unconsciously, in a practice of deceit. He is possessed of an acute mind, and has always been observant and thoughtful. He is well versed in native lore, and when a little boy herding the goats in the Lunyangwa valley witnessed the advent of Dr Laws and Mr Stuart, and the slaughter of natives conducted about that time by the Chipatulas. He carries a peculiarly happy countenance, and his merry laugh makes him a favourite with all classes. From the first his profession of Christianity has been frank and powerful. Those who know him understand how, in private discussion with the heathen, and by personal testimony always given humbly and with respect for his seniors, his influence has been wide and permanently good.

Elangeni station has always been under a native. Mawalera's brother, Makara, was placed there by Dr Steele a year before he died. It also was an experiment and justified itself. The chief of the district, Maurau, is a brother of the late Mombera, and when the school was opened in his village Makara was sent there temporarily. So great, however, was his influence, that Maurau requested that Makara should be sent to reside there permanently. Ground was given for house and gardens for the teacher, and Makara removed his family and took up his residence under Maurau, where he has conducted a most successful educational work, and where his teaching of the Word has been blessed to the conversion of many. He was the means of breaking the war-spirit in that district, and one of the first converts was the eldest son of the chief, who before was a notoriously passionate and cruel man, and ruled his slaves with an iron hand. His first act was to give his slaves their freedom, and to pay them for the work they did on his house and in his gardens. Nawambi, the ferocious war-dancer who is referred to in the chapter on William Koyi, became a new man, although not a Christian, under Makara's influence, and a school has been carried on in his district also by Makara.

Hora station, as has been stated in a former chapter, was opened by Mr and Mrs McCallum, and on their being transferred to Mwenzo station, the work there was also placed in the hands of a native teacher-evangelist who had been with Mr McCallum both at Ekwendeni and Hora. Thus, within a short space, we had given up two European workers in Ngoniland and developed native agency in the manner shown. This we look upon as a real advance, proving both the permanence of the work among the natives, and the possibility of speedily evangelising the country by means of native agents. Africa must be evangelised by the African, and although our native helpers are only moderately equipped, their work and influence serve to show that more fully trained agents obtained through the Training Institution, we may hope for greater results than we now see.

In former days the Ngoni were the troublers of their mission stations, and it is worth noticing in connection with the transfer of Mr McCallum to Mwenzo, how the Ngoni in their new character as peaceful worshippers of God were able to render assistance to a far - distant mission. Mwenzo station has lately been begun by Rev. Alexander Dewar. Houses had to be erected and the station laid out, but the natives of the district were not eager to help in such work, even although the presence of the Mission was to be for their protection and benefit. They belong to the Nyamwanga tribe which was formerly harassed by the Ngoni, many members of the tribe having been carried captive to Ngoniland. The advent of Mr and Mrs McCallum with a band of loyal Ngoni, some of whom were Church members and catechumens, was most opportune. The Ngoni were known to the Mwenzo people only as cruel warriors, constantly raiding their neighbours, but now they saw them in their midst with the implements of peaceful labour in their hands, and the Word of God in their hearts, and on their tongues. They aided in the work which the Mwenzo people declined to do, and at the same time by life and word, proclaimed the reason of the change in their manner of life. The effect of this has been very great and a valuable object-lesson in that new district, saving years of toil before the people could understand fully the meaning or fruits of the Gospel. Mr McCallum continued his teaching, and had the joy of seeing several of those who had gone with him admitted into the visible church by baptism, some time after he settled there. Thus the trial of having, for the third time, to relinquish an organized station and submit to the hardships and difficulties of pioneer work, was in some measure rewarded.

For several months in 1896 we were in considerable anxiety in connection with a threatened collision between the British Commissioner and the Ngoni. As recorded, Mombera, the paramount chief, had died, and for some years the tribe was ruled in sections by the head-men or sub-chiefs. The old desire of Ng'onomo to increase his power and attain to the chieftainship was revived. In his district, and that of Mperembe, we had made frequent efforts to be allowed to begin work but without success. We were hopeful that in all the other districts the influence of the Gospel was such that war was for ever at an end, and now our hopes were to be tested by months of turmoil and excitement. The Mission had been the only outside influence acting in Ngoniland, and no Government agent had visited the Ngoni officially since the British Consul saw Mombera in 1885 when friendly greetings were exchanged. Over all the surrounding tribes in the Protectorate, the Government had exercised its jurisdiction, and as Mperembe and Ng'onomo had not given up war and raiding, they spread the report that the Ngoni were being left alone because the Government was not strong enough to meddle with them. In this way they strove to revive the old war-spirit throughout the country, and of course our work, and especially our native helpers, came in for adverse criticism. The Evangelists in charge of districts had much to bear, but the Christians rallied to their support, and by calm and judicious behaviour they quieted many a turmoil and saved their work.

When the country was in that state, an event occurred which threatened the peace with the British Administration and gave us much trouble. The district of Kasungu, lying to the south of

Ngoniland, had been the scene of a conflict between the Commissioner's forces and the natives there. Chibisa, the chief, made his escape when his town was taken, and came to Ng'onomo as a refugee. As soon as we knew it we tried to persuade Ng'onomo to drive him away, lest trouble should come upon himself. Chibisa pretended to have a considerable army at his command, and tried to incite Ng'onomo and Mperembe to join him in an attack upon the British at Kasungu, where the latter had established a fort. Mr Swann, the Government agent at Kotakota, instead of pursuing Chibisa with an armed force, very judiciously sent policemen with a letter to us requesting Ng'onomo to hand over Chibisa to them. Ng'onomo refused. Mr Swann's intimate knowledge of African natives made him alive to the danger from the native police, who have too often overridden their commissions and become breakers rather than guardians of the peace, and the delicate business he had in hand with such a man as Ng'onomo. He had no desire to induce a rupture with Ng'onomo, and wrote: "The police have orders to obey the missionaries, and to come back if the chiefs allow Chibisa to escape or refuse to arrest him. They are in no case to do anything but visit the chiefs, take the man from them, and return." Chibisa subsequently, in alarm, fled to Mpezeni, another Ngoni chief (mentioned in the first chapter), living nine or ten days' distant to the south-west. As Mr Swann's letter was addressed to the Ngoni chiefs and required their answer, we convened meetings in different districts to discuss the situation. We pointed out to them that unless they made their position clear, they might all be involved in trouble with the Commissioner. All, save Mperembe, from the first denounced Ng'onomo for receiving Chibisa, and frequently requested us to head a war-party to go and compel him to deliver him up. The result of our meetings was, that the chiefs wrote to the Commissioner saying that since Mombera died there had been no paramount chief, and each had been simply ruling his own district as before, and could not be held responsible for Ng'onomo's conduct, as he was not under either of them, and they did not sympathise with his action, but desired to live in peace with the British. Even Mperembe began to see the inadvisability of continuing in the compact with Ng'onomo, and he also sent a letter to the Commissioner.

After Chibisa's flight messengers came from Mpezeni calling the Ngoni to rise with him and Chibisa against the British. Ng'onomo was the only chief who could be found to favour the exploit. Many old men had been eager to accept the invitation, but at the meetings held to discuss the matter, the young men, and those who had been the flower of the armies, rose as one man against the proposal to engage in war. Several hundreds were assembled, some carrying arms, but the result of the meeting was the entire defeat of the advocates of war. The young men spoke calmly and forcibly, with every respect shown to their seniors, but they were firm in their position. The young chief of Elangeni and his cousin from Ekwendeni recalled how the Gospel had come to them, and how the chiefs and head-men had for long hindered the work, until in this matter they as a tribe were left far behind other tribes. The age of war, they declared, must now be considered as dead. They said they had no desire to point the finger of scorn at their seniors, but they had apprehended a more excellent way and were to stand firm in it, and refuse to take the spear again. It was thus demonstrated to the old men that their voice was no longer a power in the tribe. As we witnessed their discomfiture, we remembered the time when some of the notables there had declared that if we got liberty to preach and teach in the tribe, we would steal the hearts of their people. The result they had feared they now witnessed. Years before this occurred, another similar incident was witnessed at Ekwendeni by Mr Stuart. On a Sunday a gathering of men was convened in the chief's village to plan a raid on a neighbouring tribe. When the hour for service came very few people turned out. The teachers went to the village and rang the bell. All the young men who had been summoned to hear the plans for war, rose up and left in a body to attend the service. The old men were left alone, and the war proposals fell to the ground in consequence.

When the excitement of the foregoing events died away, there was an increase of interest in our work. The teachers who had been persecuted were reinstated in public favour, and except for Ng'onomo we were on friendly terms with all sections of the tribe, as we now had had more cordial communications with Mperembe. The popularity of our work increased, and the services were more largely attended, while schools were desired in places where we had none. The Christian community was consolidated by means of the trial, and their influence deepened and extended.

In April 1896 Sir Harry Johnston, the Commissioner for British Central Africa, wrote to us as follows : " You will observe that in the new

Regulations extending the Hut Tax to all parts of the Protectorate, I have exempted only one district, viz., that portion of the West Nyasa District which is occupied by the Northern Ngoni. My reasons for doing so are these :—Hitherto the Ngoni chiefs have shown themselves capable of managing the affairs of their own country without compelling the interference of the Administration of the Protectorate. They have maintained a friendly attitude towards the English and have allowed us to travel and settle unhindered in and through their country. As long, therefore, as the Northern Ngoni continue this line of conduct and give us no cause for interference in their internal affairs, so long, I trust, they may remain exempt from taxation as they will put us to no expense." The Ngoni remain to this day the only tribe not under the direct jurisdiction of the British Government. They are, however, no less helpful than others in the great task of the redemption of Africa, which is now so successfully guided by the Administration of the Protectorate as the temporal head. Apart from labour given at mission stations, many hundreds of men go every year to the coffee plantations of the Shire Highlands, and to the trading corporations at work in various parts of the country, where they prove steady and successful labourers, without whom, and others, the commercial interests could not prosper. But let us not forget that all has come about through the preaching of the Gospel of Christ.

How deeply the Christian element has become fixed in the tribe was shown at the placing of a paramount chief in room of the late Mombera in 1897. When the Ngoni found themselves face to face with the British Administration, and realised the need for a chief, they proceeded to elect one. To this ceremony all sections of the tribe came. Mperembe, because he himself desired the chieftainship, had delayed the event, but he now took an active part in furthering the appointment of Mbalekelwa the eldest son of Mombera. He sent for Mawalera Tembo at Njuyu, and desired him to remain throughout the ceremony, as they did not wish to do anything wrong. It might be said that our teacher-evangelist was the most important individual there, as he was consulted on every point. He turned the occasion to good account by conducting religious worship, and subsequently addressing the assembly on the foundation of good citizenship and good government, charging the chief to rule his people by the Word of God, and for ever sheath the sword of his fathers.

But the occasion was not to pass without an attempt being made by Ng'onomo to raise the war spirit. By violent speeches and war-dances he called them to observe the customary duty in placing a new chief, viz., to send out an army "to wash their spears in blood." Mawalera and the other teachers were assaulted with opprobrious names. They stood their ground and were supported by most of those assembled, and eventually the ceremonies ended quietly and happily. Mperembe, in the dim light of his new thoughts of the Gospel, offered a sacrifice to Mombera's spirit, praying him to remember the missionaries when they taught God's Word to the people!

It was a time of great rejoicing, and one of the first acts of the chief was to signalize his accession to the throne by requesting that schools be established in his villages, and he himself desired to become a pupil of the teacher who was sent. At the same time a sub-chief in the Ekwendeni district had to be appointed in room of the late Mtwaro, who died some years before Mombera his brother. The reason for delay in this case was the confirmed objection to the son of Mtwaro, as he was a teacher in the Mission and a Church member. But God, who worked in all that was taking place, gave us the joy of seeing the people, of their own free will, choose and appoint Amon Jere, not only a Church member and teacher, but an ordained elder in the Church, to be his father's successor. At the same time Yohane Jere, the elder brother of Amon, was elected to the important office of " Father or adviser of the chiefs," so that while we never interfered with their tribal affairs, or put forward any of our pupils to positions of honour, our work was recognised in these important events.

Just before our departure on furlough in 1887, we welcomed as colleague the Rev. Donald Fraser, widely known in connection with the Student Volunteer Missionary Movement. His introduction to the work was at a time when it was at the height of the flood. The schools had been successful, the teacher-evangelists had been active in diffusing scriptural knowledge, and we had come to a reaping time. At Ekwendeni, where the Europeans were located, thirteen men, seven women, and nine children, were baptized on one Sabbath, while many more were admitted to the catechumen's class. On another Sabbath Mr Fraser went with me to Hora station, where five men, one woman, and four children were baptised, and about forty admitted as catechumens. But the blessing was not confined to stations where Europeans were working, for at Njuyu and

Mr Stuart and Ngoniland Teachers.

Elangeni, as at Hora, where our native assistants were located, there was a season of rejoicing as an abundant harvest was reaped on two other Sabbaths. I quote the account given by Mr Stuart who accompanied Mr Fraser to those places.

"On a recent Saturday we left Ekwendeni and spent the Sabbath at Elangeni. Ten young men, who had been examined at Ekwendeni and given satisfactory evidence of their faith, were admitted. As it was the first service of the kind ever held in the district, a great crowd of people turned out to witness this, to them, strange ceremony. The young son of their chief was among those baptized.

"In the afternoon we had a quiet little gathering, when sixteen of us sat down at the Lord's Table. Makara took part in this service, giving a very effective address on the love of God in sending the good news of the Gospel to the Ngoni.

"The following Sunday we spent at Njuyu. For a long time past a quiet work of grace has been going on here, the extent of it only now becoming apparent. We arrived on Friday afternoon. We saw as many as we could that night, and on Saturday we were in a state of siege nearly all day. Fifty-one persons altogether were examined, and out of these forty-four—twenty-two men and twenty-two women—were held by the Church as fit for baptism. Most of these have been for years connected with the Mission. Some of them received their first instruction in school; some are the wives of teachers, and have been taught by their husbands. One old widow is the mother of a teacher. Another woman is the mother of one of the late Dr Steele's personal boys, a boy who is now being educated at the Livingstonia Institution. The father, in this case, is still a heathen. One young man, a rescued slave, was long ago the servant of the late Mr James Sutherland. Some are husbands of heathen wives, and others wives of heathen husbands, and in not a few cases both the husband and wife were admitted. Some of them in their answers to the questions put to them showed a wonderful knowledge of divine things. Others again were, according to our ideas of knowledge, very ignorant; but they all knew that Jesus Christ died to save them, and are trusting in Him to break the power of sin in their hearts. Mawalera, whom they all look upon as their spiritual guide, had seen them all previously, they having come to him of their own accord, and he was therefore able to give us information about them. We, of course, could only see to their knowledge, and had to depend largely on the Church as to whether or not their lives were consistent with their profession. They are so closely associated together in their village life that every native knows the kind of life his neighbour leads, and one or two whom we thought right, as far as knowledge went, were rejected by the Church on account of their inconsistent lives. Polygamy and beer-drinking are the two great evils. The former and drunkenness have always excluded from membership, but total abstinence was not always a *sine qua non*. Now, however, the Church at Njuyu, having realised the dangers of example, has resolved to eschew the beer entirely. Coming thus from themselves, it is far better than if it had come from us. Reformation and not revolution will make them intelligent Christians.

"The Sabbath was a great day, and will long be remembered in the history of the Church at Njuyu. In the forenoon we had the baptismal service, and also the ordination of four elders, the first to be set apart at Njuyu for the office — Mawalera, Makara, and two others having been chosen by the Church. The choice shows the sound sense of the members.

"During the afternoon we observed the Communion, when about eighty of us partook of the Lord's Supper. The service was most impres-

sive ; simple, perhaps, there being an absence of
fine surroundings and priestly garments, and the
people, most of them, rude and ignorant.   The
worshippers, at least, were devout and earnest.
At this service Mawalera delivered a most im-
pressive address, when one woman could not
help interjecting a remark by way of emphasis.
The brotherliness existing between the . Church
members and the heartiness manifested at all the
services were marked.   Mawelera has an important
work before him in teaching these converts, but
we have confidence in him.   ' The Lord hath done
great things for us ; whereof we are glad.' "

Thus on four successive Sabbaths in Ngoniland,
eighty adults were admitted to the membership
of the Church, while several hundreds on profession of their faith were enrolled as catechumens.

Under Mr Fraser and Mr Stuart with their
native helpers, the work has been greatly extended and blessed since then.   They have
recently had the satisfaction of seeing the last
of the doors in Ngoniland opened to the work
of the Mission.   That arch-enemy of peace,
Ng'onomo, has signified his change by receiving teachers, and Mperembe too has not only
received teachers, but has become a liberal
supporter of the work by gifts of stock at
the monthly collections.   Owing to the spread-

ing out of the villages and removals to new ground, the number of the schools has been increased, so that sometimes as many as four thousand scholars are under instruction, and now school fees are being charged. Ten years before, our first school with twenty-two scholars was opened. Although that may for a time diminish the attendance, the wisdom of the step will be seen. The minimum of attainment with which we are satisfied in the case of most of those attending, is that they leave the school able to read the Word of God for themselves, and possessing a copy of it. While thousands of copies of school books have been bought by the people, there is a widespread desire to possess the Scriptures. Hundreds of Zulu Bibles, Testaments and single gospels, hymn-books and catechisms, have been sold, amounting to a large sum. The people who were wont to steal rather than work to acquire anything, now give a month's labour for a copy of the Bible, or a fortnight's for a copy of the New Testament. Men with the marks of old battles on their bodies, may be seen earnestly labouring at what once was considered ignoble work, in order to own a copy of the precious Word of God.

The liberality of the Christians is remarkable.

There is little trading with outsiders in Ngoniland, and their means of acquiring wealth, with
wages at a penny a day, are small, yet the value
of several pounds is frequently given as a church-
door collection. Besides this the people are undertaking the support of schools for their children,
and altogether it is seen to be true as Mr Fraser
wrote some months ago, "The work here has
entered on a new chapter." In another letter
he says on this point, "The monthly collections
are moving on apace. Mr Stuart and I have
turned into grain and iron merchants, and our
back-yards are huge poultry establishments.
Mr Stuart takes fowls for books, and about
120 fowls were received last week alone. The
Sunday collections are a rare sight. No less
than 150 carriers brought in the offerings from
the out-stations. My house and verandah are
packed with the produce, and the baskets, and
the hoes, and the beads which the people gave.
Two bulls, a cow, and two goats which were
contributed were bought up at once. Every
month sees a visible increase in the liberality
of the people. Then as to books. I fancy
that at this station alone quite 1000 volumes
have been sold in the last eight months."

At the Communion Service held at Ekwendeni
in May 1898, to which the following accounts

refer, the offering of the people for the service of God was as follows :—

Money, £1, 8s. 0¾d.	2 Goats.
11 Knives.	105 lbs. Beans.
14 earthenware pots.	97 lbs. Flour.
16 Baskets.	233 lbs. Maize.
1 Mat.	34 lbs. Potatoes.
67 Fowls.	62 lbs. Pumpkins.
2 Sheep.	3 lbs. 6 oz. Beads.

The total value, not including European contributions, was £3, 3s. 5¼d.

An intelligent young man of the tribe of Tonga, describing the services to a companion, said he stood at Ekwendeni and saw band after band coming over the distant ridges, and steadily marching towards the Mission station, where they were gladly received by the Christians, and taken away to the villages to be entertained. The villages were crowded with guests, men in some, women in others, and there seemed room for no more. Still, however, other bands appeared on the horizon, and as they arrived, the warmth of Christian feeling made elastic the possibilities of hospitality. "As I saw this," he said, "I marvelled." "Then the services, where, with an elder or other leading Christian, small companies gathered by themselves for prayer, and many were melted to tears—as I saw these I greatly marvelled." "Then at the baptismal service, as I saw those who were to be baptised coming forward one by one, and receiving the rite until Mr Fraser's arm grew tired, and he sat down, and Mr Henderson continued· in his place ; as I saw men with scars of spears and clubs and bullets on them ; and as I saw Impangela, the widow of Chipatula, baptized, I marvelled exceedingly. I said in my heart, 'Can these be the Ngoni submitting to God, the Ngoni who used to murder us, the Ngoni who killed the Henga, the Bisa, and other tribes ? ' And then at the Lord's Table, to see these people sitting there in the still quiet of God's presence, my heart was full of wonder at the great things God had done."

The Rev. James Henderson, of the Training Institution, has, on request, furnished the following account of the Communion Season at Ekwendeni :—

" I was travelling up the Lake shore on my way back to Livingstonia after conducting the Sacramental Services at Bandawe, when I got a letter from Mr Donald Fraser, asking me to turn aside and go up to Ekwendeni to assist him with the Sacraments there. I had left Livingstonia for the work at Bandawe on the day when the schools were closed, and now there was little more than a fortnight remaining of the vacation, and the work for the new session was all to be prepared. With the heat and the rain, the fever, the trudging through the loose sand, the constant moving from place to place, and especially after the excitement and strain of the last great week at Bandawe, I was as tired as could be, and my inclination was to pass on and choose the first deserted bay for a camp, where, for a few days, I might cease to be a missionary or any other sort of thinking animal. But Saturday afternoon found us striking inland across the mountains for Ekwendeni, and spending Sunday among the Tumbuka people on the uplands. I reached the station on Monday evening. I can never be too thankful that I did not miss seeing, and that I was privileged to take some small part in, this remarkable sacramental gathering.

" 'The meetings were to begin in the week after I arrived. Mr Fraser had decided to have things very much on the general lines they used to follow at a Communion Season in the Scottish Highlands. The examination of catechumens seeking to go forward to baptism was finished, and it was known how unprecedentedly large was the number to be admitted. It was felt on all hands that this would be a great opportunity for reaching the hearts of those that had hitherto held aloof. For some time the native Christians had been assembling daily for prayer, and now they were looking forward to a time of great spiritual awakening and quickening. The missionaries were expecting more than that. It would be a gathering of the whole Christian Church of the tribe, and the missionaries trusted that the Church as a body might be led to make a further forward step in spiritual experience. They looked away to the untouched regions beyond, and they were praying for an outpouring of the Spirit that there might be fuller consecration to the Lord, and new devotion and enthusiasm for His service.

"Of course there was no building at all adequate for the expected congregations. A temporarary open-air church had to be set up. That is not difficult in Africa. Posts, split branches, and the strong, tall grass, provided a screened enclosure—protection from the wind is what is required—and some bricks and a few planks make all that is needed for a platform. Seats can be dispensed with on such occasions. They are ornamental rather than necessary.

"On Monday the people began to assemble. The first - comers were from Mperembe's, the raiding chief who had only very lately received teachers. They had brought a contribution in the shape of a sheep and a goat from the chief himself, and their appearance was a picture of Mperembe's attitude to the Mission. They were evidently intended to make up by numbers what they lacked in age and status, and were altogether a non-committal deputation, raw and rustic, and a good deal 'out of it' when the other peoples gathered. By Tuesday evening the footpaths were full. Whole families were coming, the mothers and daughters carrying cooking-pots on their heads and bags of flour, the men with strings of maize cobs on their shoulders and other produce of their gardens for the 'collection,' and often a tired child on their backs. Most of them were dressed in snowy white calico. They travelled silently; and the people in the heathen villages by the way climbed up the ant-hills to look at them, and called, so they told us, 'What has happened? What *impi* is after you?' Past the old houses of the first Ekwendeni station, across the rising ground, we could see them coming in a long straggling Indian file, which changed into solid masses as they crossed the river and came up the Mission road. It was then that I realised the nature of the work that was being done among the tribe. The swinging pace could not be mistaken, even before the individuals could be seen. It was the fighting men, the men in the prime of their strength that the Gospel had laid hold of. A fitter-looking set of men and women it would be hard to find anywhere. What a promise they are of the speedy coming of the Kingdom of Christ in that land. All Wednesday forenoon they streamed in, people that had come from far, and slept one or two nights on the way.

"The enclosure was intended to hold somewhat over a thousand, but it was soon evident that an extension of it had to be made. The 'hospitality committee,' the directors of which were the two local chiefs, Yohane and Amon, had their resources taxed to the utmost. Every hut in the neighbouring villages was taken over by them, and when these were filled they made use of the cattle kraals. There was no trouble made about it. I heard that when Yohane was asked first how many people he could accommodate, he thought he might manage with a hundred, but when the need arose he himself arranged for over a thousand.

"On Wednesday at mid-day the first meeting was held. Well on to 3000 people were present. They had taken their places many of them long before the hour, and when we went down they were singing a hymn. Exuberance of spirits is the characteristic of this Church, and the singing is something always to be remembered, but the meeting had not gone far on before it became clear that it was not the familiar mood of the people that had to be dealt with. They were so bent on hearing that they altogether forgot their correct listening manners, somewhat ostentatious in attention, and sat up, looking straight at the speaker. They had evidently made up their minds that they were to learn something. The solemnity of the consciousness of the presence of the Lord seemed to creep over the whole assembly. I confess that as I sat and listened, while Mr Fraser and Mr Stuart addressed the people, touching them and swaying them with their words, something like fear came over me, and ·I doubted whereunto this would grow. Would excitement seize the people, and could it be controlled? That day I prayed far more that no evil might befall the gathering than that good might come. But my fear was foolish. The thing was of the Lord. It was in His hands.

" Most of the addresses were intended for the believers, and dealt much with heart-sinfulness. The touch-stone of self-examination proposed was conscious daily communion with the living Christ. The addresses were pointedly practical, calling for definite acts of self-surrender to the Lord. Men and women were entreated to deal with their Saviour about the things which they found standing between Him and them. The person and work of the Holy Spirit were brought prominently forward.

" The opening day, as might be expected, was one of perplexity. The Christians had thought, perhaps, that the meetings were intended almost wholly for those outside, and they found that it was themselves that were addressed. Their self-examination brought with it sorrow and humiliation. The meetings after that took a new tone. There was more stillness and less self-certainty. The Christians were seeking God again, and striving with the chains of self and the world.

" One night the teachers asked Mr Fraser and myself to meet with them in the house where they were staying to explain to them difficulties, mostly with reference to the work of the Spirit, which were troubling them. Mr Fraser had been with them alone the previous night. We sat on the ground along the walls, dim fires burning on the floor down the middle of the long building. There was no formal teaching or speaking. In the semi-darkness the men talked with the utmost frankness ; and there was a time of confession and prayer. Several of the older men gave signs of having been brought very near to God. Coming outside we found the air full of hymns. A tropical moon was full in the sky, and in the villages all around the people were met out of doors singing, and as the soft breeze rose and fell the words were borne to our ears. The Evening Prayers were long over; but the hearts of the people were too full for rest.

"In the middle of Friday night an incident occurred which might have been attended with evil consequences among a people naturally superstitious. A little child, brought by its parents to be baptised, they themselves also being at the same time admitted into the Church, was taken ill; the journey had been too much for it. They came with it to us for treatment, when it was far too late, and as we were looking at it, it died. It was a heavy trial to them. I think it was their only child. They were away from home and none of the usual customs of honouring the dead could be observed. But they took the situation bravely. They made no noisy lamentation. All night they sat alone beside the little body in one of the Mission rooms, and, when the morning Prayer Meeting was over, with a company of friends they carried it forth to burial. It was laid in the little plot where the remains of Dr Steele lie at rest. And the great congregation looked down the slope, and many of them wondered for they were seeing a Christian burial for the first time. We were singing a hymn. There was no wild weeping and wailing. And from the side of the grave were

borne to them the triumphant words in which
the Christian proclaims his faith in the resur-
rection, 'Death is swallowed up in victory.
O death where is thy sting?   O grave where
is thy victory? . . . Thanks be to God which
giveth us the victory through our Lord Jesus
Christ.'   God made death even to work for our
good.   It was a lesson without which the other
teaching would have hardly been complete.

"The second service of Saturday was the
meeting in which most interest centred.   The
congregation, which had been growing larger
every day, was now almost at its greatest.
Every corner of the enclosure was filled.   In
the centre sat the men and women to be baptised.   The native elders arranged them according to their districts.   They were in all 195
adults.   Again I was struck with their appearance.   They were the pick of the people in the
prime of life.   Such a sight was never before
seen among us, and rarely in the whole history
of the spread of our Faith.   Oftener than once
the head that was bowed before us, as they
came forward to receive the rite, was scarred
with an old wound; for several of those baptised had taken part in the very last raid of
this section of their tribe.   It was a wonderful
day of ingathering that we were privileged to
see, and we were but lately come into the field.

Some of the early sowers could only hear tell of it, and others—Koyi, Sutherland and Steele —were with their God. The day was of the Lord. We were as onlookers upon His doings. It was natural that He should do great things.

"The celebration of the Lord's Supper was held upon the Sabbath forenoon, when the congregation amounted to nearly 4000. The enclosure was packed to its utmost capacity. On a large ant-hill outside a crowd of curious sight-seers assembled, grey-bearded old men, Zulu *ndunas* with the ring of hair crowns, and skin - clad heathen from the remoter villages, and around the fence stood hundreds of miserable - looking Tumbuka women, craning red-ochred heads over the grass to watch the proceedings. The addresses were calls to action. If the Lord had done any-thing for His people in the course of the meetings they were now to let it appear in the vows which they renewed before Him. Seated in rows in front and on each side of the platform the Church members formed a large body, but the greater mass of the people was still beyond the separating space. What was true of this assembly was true, on a greater scale of disproportion, of the land. The Gospel had as yet come only to the few, and the few must bear it to the many. The King was present at His feast, and He was enter-ing into the full possession of many subjects. In

His presence anxiety was giving way to peace. In His keeping the future could be faced with joy. Never was such a collection taken in the land before. A heap of Indian corn nearly breast high all but blocked the side entrances. They had given of all kinds of their possessions, money, beads, rings, bracelets, knives, hoes, axes, pots, baskets, mats, pumpkins, millet, beans, potatoes, fowls, sheep, goats, and cattle.

" In the afternoon the children were brought for baptism, 89 in all, thus making a total of 284 souls received at the one time into the Church.

" The feast was now over; the time for work had come. At dawn the next morning we met for the last time to give thanks. By mid-day we were scattered along every outward path, and some knew that they were not, and might never more, be alone, for they had learned to walk in conscious fellowship with the Comforter, the Holy Spirit."

" It is the Lord's doing and is wondrous in our eyes.

" Amen : Blessing, and glory, and wisdom, and thanksgiving, and honour, and power, and might, be unto our God for ever and ever. Amen."

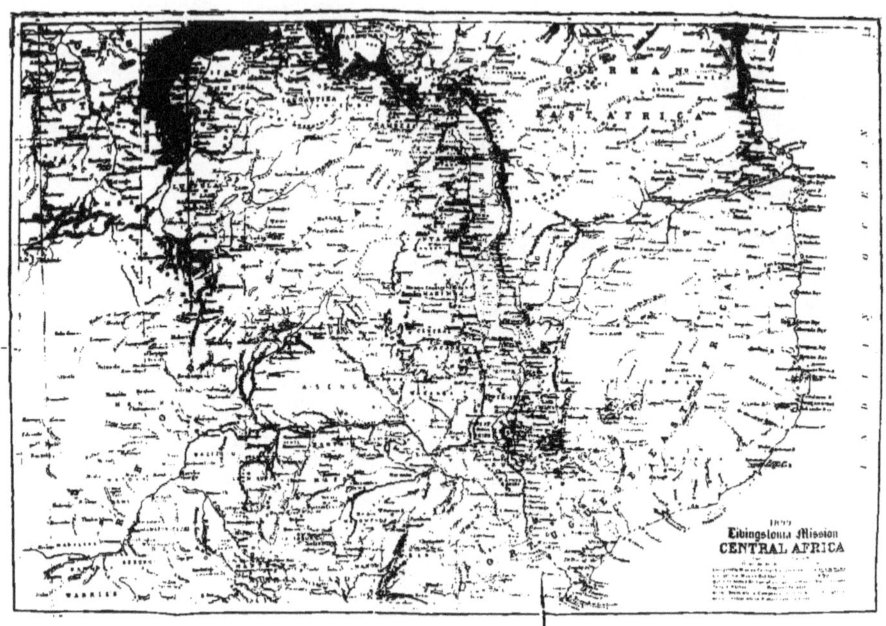

GERMAN
EAST AFRICA
INDIAN OCEAN
Livingstonia Mission
CENTRAL AFRICA

INDEX

THE LIVINGSTONIA MISSION.

MISSIONARIES FROM SCOTLAND.

(There are 300 Native Christian Teachers and Students.)

LIVINGSTONIA TRAINING INSTITUTION.

Rev. ROBERT LAWS, D.D., M.D. (Aber.), F.R.G.S., and Mrs LAWS.

Rev. JAMES HENDERSON, M.A.(Ed.)

Miss L. A. STEWART, Teacher.

Miss M. JACKSON, Nurse.

Miss M. M'CALLUM, Nurse and Teacher.

Mr W. MURRAY, Carpenter.

Mr W. THOMSON, Printer, and Mrs THOMSON.

Mr W. D. MACGREGOR, Carpenter, and Mrs MACGREGOR.

Mr M. MOFFAT, Agriculturist.

Mr W. J. HENDERSON, Artisan Evangelist.

KARONGA.

Mr ROBERT SCOTT, M.B., Ch.M. (Glas.)

Dr J. M. HENDERSON, Teacher.

NGONILAND—EKWENDENI.

Dr W. A. ELMSLIE, M.B., C.M.(Aber.), F.R.G.S., and Mrs ELMSLIE.

Rev. DONALD FRASER.

Mr CHAS. STUART, Teacher.

BANDAWE.

Rev. A. G. MACALPINE, and Mrs MACALPINE.

Dr GEO. PRENTICE, L.R.C.P. & S. (Edin.)

Mr R. D. MACMINN, Teacher.

MWENZO.

Rev. ALEX. DEWAR, F.R.G.S., and Mrs DEWAR.

Mr PETER McCALLUM, Artisan Evangelist, and Mrs McCALLUM.

COMMITTEE. *(Meeting in Glasgow.)*

From Foreign Missions Committee.

RIGHT HON. LORD OVERTOUN, 7 W. George St., Glasgow, *Convener.*
ROBERT M'CLURE, ESQ., 145 St Vincent Street, Glasgow, *Secretary.*

Rev, ALEX. ALEXANDER, M.A., Dundee.
Mr W. L. BROWN, B.L., Glasgow.
Colonel T. CADELL, V.C., Edinburgh.
Sir JOHN COWAN, Bart., Beeslack.
Rev. J. FAIRLEY DALY, B.D., Glasgow.
Mr WALTER DUNCAN, Glasgow.
Mr W. FERGUSON, LL.D., Kinmundy.
Rev. R. FORGAN, B.D., Rothesay.
Sir WILLIAM HENDERSON, LL.D., Aberdeen.
Rev. T. B. KILPATRICK, B.D., Aberdeen.
Rev. Prof. LINDSAY, D.D., F.R.S.E., Glasgow.

Dr LOUDON, Hamilton.
Mr D. MACLEAN, Glasgow.
Rev. ALEX. MILLER, B.D., Buckie.
Dr JOHN MOIR, Edinburgh.
Rev. Prof. ROBERTSON, D.D., Aberdeen.
Mr JOSEPH C. ROBERTSON, Glasgow.
Mr JOSEPH RUSSELL, Port-Glasgow.
Rev. R. J. SANDEMAN, D.D., Edinburgh.
Dr GEORGE SMITH, C.I.E., F.R.G.S., Edinburgh.
Rev. A. SOUTAR, M.A., Thurso.
Rev. J. STALKER, D.D., Glasgow.
Mr JOHN STEPHEN, Glasgow.
Ordained Missionaries on furlough.

From Glasgow.

Mr GILBERT BEITH, Glasgow.
Mr THOMAS BINNIE, Glasgow.
Mr ROBERT BRODIE, Glasgow.
Mr HUGH BROWN, Glasgow.
Rev. JAMES BROWN, M.A., Glasgow.
Mr JAMES CAMPBELL, Tullichewan.
Rev. JOSEPH CORBETT, D.D., Glasgow.
Mr HUGH DAVIDSON, Lanark.
Rev. R. HILL, M.A., Renfrew.
Rev. R. HOWIE, M.A., Govan.
Mr CHARLES C. DUNCAN, Dundee.
Mr ANDREW HUTCHESON, Perth.
Mr T. R. JOHNSTONE, Glasgow.
Mr ROBERT M'CLURE, *Secretary*, Glasgow.

Rev. A. R. MacEWEN, D.D., Glasgow.
Rev. JAMES MILLER, Bridge-of-Allan.
Mr STODART J. MITCHELL, Aberdeen.
Mr FRED. L. M. MOIR, Glasgow.
Rev. G. REITH, D.D., Glasgow.
Mr A. ELLISON ROSS, S.S.C., *Treasurer*, Edinburgh.
Mr A. SOMERVILLE, B.Sc., F.L.S., Glasgow.
Mr JAMES STEVENSON, F.R.G.S., Largs.
Mr JAMES THIN, Edinburgh.
Rev. JAMES WELLS, D.D., Glasgow.

General Treasurer.
ALEXANDER ELLISON ROSS, ESQ.,
FREE CHURCH OFFICES, EDINBURGH.